Emerging Pattern of Educational Technology

ABOUT THE AUTHOR

Sandra Roldan, is and Assistant professor and Research coordinator of College of Teacher Education at Laguna State Polytechnic University, Philippines. She finihed his bachelor of secondary Education major in General Science and Masters of Arts in Teaching major in Science and Technology and has completed her academics requirement in Doctor of Philosophy in Education major in Educational Leardership and Management. She is teaching major subjects both in the undergraduate and graduate programs. She has presented her papers in national and international conferences and published in international journals.

ABOUT THE BOOK

Technology education is a study of technology, which provides an opportunity for students to learn about the processes and knowledge related to technology. As a study, it covers the human ability to shape and change the physical world to meet needs, by manipulating materials and tools with techniques. As education, the goal is to teach the knowledge and techniques to develop technological literacy which is accomplished by bringing laboratory activities to students. Technology education is an innovative approach to technology education in Australia, which differs in that it is accomplished by situating students in the context of the need. The term "technology education" is frequently shortened to "tech ed". Educational technology is most simply and comfortably defined as an array of tools that might prove helpful in advancing student learning. Educational Technology relies on a broad definition of the word "technology". Technology can refer to material objects of use to humanity, such as machines or hardware, but it can also encompass broader themes, including systems, methods of organization, and techniques. Some modern tools include but are not limited to overhead projectors, laptop computers, and calculators. Newer tools such as "smartphones" and games (both online and offline) are beginning to draw serious attention for their learning potential. This book is provides a thorough examination of the use of advanced educational technology as well as point out to the new directions and its applications in the imparting of education at various levels.

Emerging Pattern of Educational Technology

Sandra Roldan

AMIGA PRESS. INC
Delhi -110051

AMIGA PRESS. INC
H.O :- 28-A, Indra Park, Chander Nagar
Delhi :- 110051 (India)
Tel :- 011-22504028, +91-98736-57000
Email :- info@amigapress.co.in
amigapub@gmail.com
Website :- www.amigapress@co.in

This edition is published by Amiga Press.Inc

ISBN : 978-93-87295-08-7

First Published 2019

Printed and bound in India by Replika Press Pvt. Ltd.

Preface

Educational technology is the study and ethical practice of facilitating learning and improving performance by creating, using and managing appropriate technological processes and resources. The term educational technology is often associated with, and encompasses, instructional theory and learning theory. While instructional technology is "the theory and practice of design, development, utilisation, management, and evaluation of processes and resources for learning," according to the Association for Educational Communications and Technology (AECT) Definitions and Terminology Committee, educational technology includes other systems used in the process of developing human capability. Educational technology includes, but is not limited to, software, hardware, as well as Internet applications, such as wikis and blogs, and activities. But there is still debate on what these terms mean.

Constructivist learning environments require students to utilize their prior knowledge and experiences to formulate new, related, and/or adaptive concepts in learning. Under this framework the role of the teacher becomes that of a facilitator, providing guidance so that learners can construct their own knowledge. Constructivist educators must make sure that the prior learning experiences are appropriate and related to the concepts being taught. Jonassen (1997) suggests "well-structured" learning environments are useful for novice learners and that "ill-structured" environments are only useful for more advanced learners. Educators utilizing technology when teaching with a constructivist perspective should choose technologies that reinforce prior learning perhaps in a problem-solving environment.

Educational technology is most simply and comfortably defined as an array of tools that might prove helpful in advancing student learning. Educational Technology relies on a broad definition of the word "technology". Technology can refer to material objects of use to humanity, such as machines or hardware, but it can also encompass broader themes, including systems, methods of organization, and techniques. Some modern tools include but are not limited to overhead projectors, laptop computers, and calculators. Newer tools such as "smartphones" and games (both online and offline) are beginning to draw serious attention for their learning potential.

Technology education allows learners to explore a variety of activities related to many areas of human endeavor. Learners can develop problem solving strategies and work habits that will be useful in almost any career and or occupation. Learners should develop a greater appreciation for the work of craft workers and the skill required of that work. Within the scope of this educational program is the effort to develop "soft skills" within the learner, as well as an opportunity for the learner to see how systems work together and the chance to put much of the academic class instruction to work in an applied way.

Technology is rapidly changing in all spheres of life, so in education. It has become very necessary to know how we learn, how we express ourselves, and how we perceive and interact with our world. As a means of responding to the change and surviving in an increasingly competitive educational environment, management institutes must develop a strategic plan for enhancing management education and value adding to the existing system of education. As a means of achieving the aforesaid goal the use of computers and other technology in support of student learning, accommodating the full range of learning styles and special needs, and fully integrating technology into the curriculum has become extremely needful in this hour. A lot of related issues are also involved in such a process as this includes administrative, planning and budgeting; professional development for teachers and administrators; and, very importantly, ensuring equity of access and use. The current emphasis is ensuring that technology is used effectively to create new opportunities for learning and to promote student achievement.

This is a reference book. All the matter is just compiled and edited in nature, taken from the various sources which are in public domain.

This book is provides a thorough examination of the use of advanced educational technology as well as point out to the new directions and its applications in the imparting of education at various levels.

—*Editor*

Contents

1

Perspectives and Meaning of Education Technology

Educational technology is most simply and comfortably defined as an array of tools that might prove helpful in advancing student learning. Educational Technology relies on a broad definition of the word "technology". Technology can refer to material objects of use to humanity, such as machines or hardware, but it can also encompass broader themes, including systems, methods of organization, and techniques. Some modern tools include but are not limited to overhead projectors, laptop computers, and calculators. Newer tools such as "smartphones" and games (both online and offline) are beginning to draw serious attention for their learning potential.

Those who employ educational technologies to explore ideas and communicate meaning are learners or teachers. The word technology for the sister fields of Educational and Human Performance Technology means "applied science." In other words, any valid and reliable process or procedure that is derived from basic research using the "scientific method" is considered a "technology."

Educational or Human Performance Technology may be based purely on algorithmic or heuristic processes, but neither necessarily implies physical technology. The word technology, comes from the Greek "Techne" which means craft or art. Another word "technique", with the same origin, also may be used when considering the field Educational technology. So Educational technology may be extended to include the techniques of the educator.

A classic example of an Educational Psychology text is Bloom's 1956 book, *Taxonomy of Educational Objectives*. Bloom's taxonomy is helpful when designing learning activities to keep in mind what is expected of—and what are the learning goals for— learners. However, Bloom's work does not explicitly deal with educational technology *per se* and is more concerned with pedagogical strategies.

According to some, an Educational Technologist is someone who transforms basic educational and psychological research into an evidence-based applied science (or a technology) of learning or instruction. Educational Technologists typically have a graduate degree (Master's, Doctorate, Ph.D., or D.Phil.) in a field related to educational psychology, educational media, experimental psychology, cognitive psychology or, more purely, in the fields of Educational, Instructional or Human Performance Technology or Instructional (Systems) Design. But few of those listed below as theorists would ever use the term "educational technologist" as a term to describe themselves, preferring terms like "educator". The transformation of educational technology from a cottage industry to a profession is discussed by Shurville, Browne, and Whitaker.

History

For several decades, vendors of equipment such as laptop computers and interactive white boards have been claiming that their technologies would transform classrooms and learning in many positive ways, but there has been little evidence provided to substantiate these claims. To some extent, the history of educational technology has been marked by a succession of innovations that arrive with much fanfare but often fade into the background once fully tested, as Cuban argues in the above title.

Theories and Practices

Three main theoretical schools or philosophical frameworks have been present in the educational technology literature. These are Behaviourism, Cognitivism and Constructivism. Each of these schools of thought are still present in today's literature but have evolved as the Psychology literature has evolved.

Behaviourism

This theoretical framework was developed in the early 20th century with the animal learning experiments of Ivan Pavlov, Edward Thorndike, Edward C. Tolman, Clark L. Hull, B.F. Skinner and many others. Many psychologists used these theories to describe and experiment with human learning. While still very useful this philosophy of learning has lost favour with many educators.

Skinner's Contributions

B.F. Skinner wrote extensively on improvements of teaching based on his functional analysis of Verbal Behaviour, and wrote "The Technology of Teaching", an attempt to dispel the myths underlying contemporary education, as well as promote his system he called programmed instruction. Ogden

Lindsley also developed the Celeration learning system similarly based on behaviour analysis but quite different from Keller's and Skinner's models.

Cognitivism

Cognitive science has changed how educators view learning. Since the very early beginning of the Cognitive Revolution of the 1960s and 1970s, learning theory has undergone a great deal of change. Much of the empirical framework of Behaviourism was retained even though a new paradigm had begun. Cognitive theories look beyond behaviour to explain brain-based learning. Cognitivists consider how human memory works to promote learning.

After memory theories like the Atkinson-Shiffrin memory model and Baddeley's Working memory model were established as a theoretical framework in Cognitive Psychology, new cognitive frameworks of learning began to emerge during the 1970s, 80s, and 90s. It is important to note that Computer Science and Information Technology have had a major influence on Cognitive Science theory. The Cognitive concepts of working memory (formerly known as short term memory) and long term memory have been facilitated by research and technology from the field of Computer Science. Another major influence on the field of Cognitive Science is Noam Chomsky. Today researchers are concentrating on topics like Cognitive load and Information Processing Theory.

Constructivism

Constructivism is a learning theory or educational philosophy that many educators began to consider in the 1990s. One of the primary tenets of this philosophy is that learners construct their own meaning from new information, as they interact with reality or others with different perspectives.

Constructivist learning environments require students to utilize their prior knowledge and experiences to formulate new, related, and/or adaptive concepts in learning. Under this framework the role of the teacher becomes that of a facilitator, providing guidance so that learners can construct their own knowledge. Constructivist educators must make sure that the prior learning experiences are appropriate and related to the concepts being taught. Jonassen (1997) suggests "well-structured" learning environments are useful for novice learners and that "ill-structured" environments are only useful for more advanced learners. Educators utilizing technology when teaching with a constructivist perspective should choose technologies that reinforce prior learning perhaps in a problem-solving environment.

Connectivism

Connectivism is "a learning theory for the digital age," and has been developed by George Siemens and Stephen Downes based on their analysis of the limitations of behaviourism, cognitivism and constructivism to explain the effect technology has had on how we live, how we communicate, and how we learn. Donald G. Perrin, Executive Editor of the International Journal of Instructional Technology and Distance Learning says the theory "combines relevant elements of many learning theories, social structures, and technology to create a powerful theoretical construct for learning in the digital age."

OBJECTIVES OF TECHNOLOGY EDUCATION

Technology education allows learners to explore a variety of activities related to many areas of human endeavor. Learners can develop problem solving strategies and work habits that will be useful in almost any career and or occupation. Learners should develop a greater appreciation for the work of craft workers and the skill required of that work. Within the scope of this educational program is the effort to develop "soft skills" within the learner, as well as an opportunity for the learner to see how systems work together and the chance to put much of the academic class instruction to work in an applied way.

Many of the skills that are to be taught in technology education classes are identified in the 1991 SCANS Secretary's Commission on Achieving Necessary Skills Report. Many of these soft skills may be better developed in a lab, problem solving setting than in a typical academic classroom.

Technology Education versus Technical Education

The goal of technology education is different from technical education. Technical education teaches specific skills directed toward a specific type of work, for example, carpentry, automotive maintenance, welding, or computer network administration, whereas technology education has a more generalized approach. Technology education is often confused with the term educational technology, which denotes technological devices and methods of delivering or developing educational information.

Technology in Education

Many people warn of the possible harmful effects of using technology in the classroom. Will children lose their ability to relate to other human beings? Will they become dependent on technology to learn? Will they find inappropriate materials? The same was probably said with the invention of the printing press, radio, and television.

All of these can be used inappropriately, but all of them have given humanity unbounded access to information which can be turned into knowledge. Appropriately used— interactively and with guidance— they have become tools for the development of higher order thinking skills.

Inappropriately used in the classroom, technology can be used to perpetuate old models of teaching and learning. Students can be "plugged into computers" to do drill and practice that is not so different from workbooks. Teachers can use multimedia technology to give more colourful, stimulating lectures. Both of these have their place, but such use does not begin to tap the power of these new tools.

In this area, you will find descriptions of how computers can be used to stimulate and develop writing skills, collaborate with peers in foreign countries, do authentic kinds of research that is valuable to the adult world, and do complex kinds of problem solving that would otherwise be impossible.

HISTORY OF EDUCATIONAL TECHNOLOGY

Educational technology in a way could be traced back to the emergence of very early tools, *e.g.*, paintings on cave walls. But usually its history starts with educational film or Sidney Pressey's mechanical teaching machines in the 1920s.

The first large scale usage of new technologies can be traced to US WWII training of soldiers through training films and other mediated materials. Today, presentation-based technology, based on the idea that people can learn through aural and visual reception, exists in many forms, *e.g.*, streaming audio and video, or PowerPoint presentations with voice-over. Another interesting invention of the 1940s was hypertext, i.e., V. Bush's memex.

The 1950s led to two major, still popular designs. Skinners work led to "programmed instruction" focusing on the formulation of behavioural objectives, breaking instructional content into small units and rewarding correct responses early and often. Advocating a mastery approach to learning based on his taxonomy of intellectual behaviours, Bloom endorsed instructional techniques that varied both instruction and time according to learner requirements.

Models based on these designs were usually referred to as computer-based training", Computer-aided instruction or computer-assisted instruction in the 1970s through the 1990s. In a more simplified form they correspond to today's "e-contents" that often form the core of "e-learning" set-ups, sometimes also referred to as web-based training or e-instruction. The course designer divides learning contents into smaller chunks of text augmented with graphics and multimedia presentation. Frequent Multiple Choice questions with immediate feedback are added for self-assessment and guidance. Such e-contents can rely on standards defined by IMS, ADL/SCORM and IEEE.

The 1980s and 1990s produced a variety of schools that can be put under the umbrella of the label Computer-based learning. Frequently based on constructivist and cognitivist learning theories, these environments focused on teaching both abstract and domain-specific problem solving. Preferred technologies were micro-worlds, simulations and hypertext.

Digitized communication and networking in education started in the mid 80s and became popular by the mid-90's, in particular through the World-Wide Web, e-Mail and Forums. There is a difference between two major forms of online learning. The earlier type, based on either Computer Based Training or Computer-based learning, focused on the interaction between the student and computer drills plus tutorials on one hand or micro-worlds and simulations on the other.

Both can be delivered today over the WWW. Today, the prevailing paradigm in the regular school system is Computer-mediated communication, where the primary form of interaction is between students and instructors, mediated by the computer. CBT/CBL usually means individualized learning, while CMC involves teacher/tutor facilitation and requires scenarization of flexible learning activities. In addition, modern ICT provides education with tools for sustaining learning communities and associated knowledge management tasks. It also provides tools for student and curriculum management.

In addition to classroom enhancement, learning technologies also play a major role in full-time distance teaching. While most quality offers still rely on paper, videos and occasional CBT/CBL materials, there is increased use of e-tutoring through forums, instant messaging, video-conferencing etc. Courses addressed to smaller groups frequently use blended or hybrid designs that mix presence courses with distance activities and use various pedagogical styles. The 2000s emergence of multiple mobile and ubiquitous technologies gave a new impulse to situated learning theories favouring learning-in-context scenarios. Some literature uses the concept of integrated learning to describe blended learning scenarios that integrate both school and authentic settings.

CRITICISM OF EDUCATIONAL TECHNOLOGY

Although technology in the classroom does have many benefits, there are clear drawbacks as well. Limited access to sufficient quantities of a technology, lack of training, the extra time required for the implementations of technology, and the apprehension associated with assessing the effectiveness of technology in the classroom are just a few of the reasons that technology is often not used extensively in the classroom. To understand educational technology one must also understand theories in human behaviour as behaviour is affected by technology.

Media Psychology is the study of media, technology and how and why individuals, groups and societies behave the way they do. The first Ph.D programme with a concentration in media psychology was started in 2002 at Fielding Graduate University by Bernard Luskin. The Media Psychology division of APA, division 46 has a focus on media psychology. Media and the family is another emerging area affected by rapidly changing educational technology.

Digital Divide

One of the greatest barriers of integrating technology into the school system deals with the digital divide. The concept of the digital divide was originally defined as a gap between those who have access to digital technologies and those who do not. This access is associated with age, gender, education, income, ethnicity, and geography. The first deals with the onset of integrating technology into the curriculum and the gap between the digital haves and have nots. In most cases, this form of the digital divide means that those who have access to a computer and the Internet are considered a digital have, while on the other hand, those who do not are considered a digital have not. In today's society, this is still a significant barrier to implementing technology into the curriculum because the socio-economic status of a school, and its students, will impact whether resources can be purchased and implemented in the school system. Schools that are able to provide technology within the classroom are able to expose their students to a new means of learning, while the students in lower socio-economic schools may miss out on these experiences.

As more and more people have gone online and started using the Internet for an increasing number of activities, researchers have begun to reconsider the notion of the digital divide. Some scholars offered a redefined understanding by seeing the digital divide as a complex and dynamic phenomenon that is essentially multifaced and includes technical access (the physical availability of technology) and social access (the mix of professional knowledge, economic resources, and technical skills required for effectual use of echnology). This means that even if schools and students have access to technology, the ways in which teachers use and introduce it is significant to consider. This form of the digital divide is yet another barrier because it also goes hand-in-hand with the resources the schools have and the training teachers receive. If a teacher, for example, is not well equipped and confident in utilising a form of technology, those students will miss out on gaining the valuable skills required for today's society.

Another factor that plays into the digital divide, which makes it difficult to implement technology into the curriculum, is the generational digital divide.

Herrington recognises that the generational divide is interpreted to mean that people on one side of the gap, including the youth, have more access and a greater ability to use new technologies than those on the other side like the adults who were born before the advent of the Internet. The generational digital divide is a common barrier because it challenges teachers to keep up with the ever-changing technology in the classroom.

Even extending beyond the classroom, by the time an individual "adopts a technology, a new one is developed, marketed, and requires a new adoption cycle". Students, who have grown up in a digital environment, may be well acquainted with the on-going process of new technological innovation but may be lacking the guidance they need in order to utilise these technologies effectively. From the teacher's perspective, this process could be an intimidating experience because something as foreign as the computer and Internet must first be learned and then taught to the students in a classroom setting. It is difficult to formulate a curriculum, which aims to integrate technology into the classroom, when the decision-makers are still in the process of learning about it themselves.

Teacher Training

Similar to learning a new task or trade, special training is vital to ensuring the effective integration of classroom technology. The current school curriculum tends to guide teachers in training students to be autonomous problem solvers. This has become a significant barrier to effective training because the traditional methods of teaching have clashed with what is expected in the present workplace. Today's students in the workplace are increasingly being asked to work in teams, drawing on different sets of expertise, and collaborating to solve problem. These experiences are not highly centered on in the traditional classroom, but are twenty-first century skills that can be attained through the incorporation and engagement with technology. Changes in instruction and use of technology can also promote a higher level of learning among students with different types of intelligence. Therefore since technology is not the end goal of education, but rather a means by which it can be accomplished, educators must have a good grasp of the technology being used and its advantages over more traditional methods. If there is a lack in either of these areas, technology will be seen as a hindrance and not a benefit to the goals of teaching.

Another major issue arises because of the evolving nature of technology. Teachers may find themselves acting as perpetual novices when it comes to learning about technology. This is because technology, including the Internet and its range of applications, is always in a state of change and teachers must attempt to keep current. The ways in which teachers are taught to use technology is also outdated because the primary focus of training is on computer literacy, rather than the deeper, more essential understanding and mastery of technology for information processing, communication, and problem solving. New resources have to be designed and distributed whenever the technological platform has been changed. However, finding quality materials to support classroom objectives after such changes is often difficult even after they exist in sufficient quantity and teachers must design these resources on their own. The study by Harris notes that the use of random Professional Development days is not adequate enough in order to foster the much-needed skills required to teach and apply technology in the classroom. Learning, therefore, becomes and on-going process, which takes time and a strong commitment among the community of educators.

Teacher training faces another drawback when it comes to one's mindset on the integration of technology into the curriculum. The generational divide might also lead to a generational bias, whereby teachers do not feel the need to change the traditional education system because it has been successful in the past. This does not necessarily mean it is the right way to teach for the current and future generations. Considering the fact that today's students are constantly exposed to the impacts of the digital era, learning styles, and the methods of collecting information has evolved. To illustrate this concept Jenkins states, "students often feel locked out of the worlds described in their textbooks through the depersonalised and abstract prose used to describe them," whereas games can construct worlds for players to move through and have some stake in the events unfolding. Even though technology can provide a more personalised, yet collaborative, and creative, yet informative, approach to learning, it may be difficult to motivate the use of these contemporary approaches among teachers who have been in the field for a number of years.

Assessment

Research has shown that there is a great deal of apprehension associated with assessing the effectiveness of technology in the classroom and its development of information-age skills. This is because information-age skills, also commonly referred to as twenty-first century literacies, are relatively new to the field of education. These include "the set of abilities and skills where aural, visual, and digital literacy overlap". Jenkins modifies this definition by acknowledging them as building on the foundation of traditional literacy, research skills, technical skills and critical-analysis skills taught in the classroom. Current school assessments are based on standardised tests and the ability to complete these uniform tests, regardless of one's preferred learning style. Many factors play into this observation including the strong impact of time. By using technology and learning through discovery, teachers may feel that they are not able to cover the material needed to meet the requirements of the curriculum. Therefore, the traditional style of teaching, including the lecturing in front of the class, and a "one-size-fits-all" approach to testing is common in today's classrooms. This is a barrier because it prevents the full integration of technology into the curriculum, the ability to learn through enquiry, and the collaborative problem-solving skills, which prove to be essential traits needed in the twenty-first century.

MEANING OF EDUCATIONAL TECHNOLOGY

Educational technology is the study and ethical practice of facilitating learning and improving performance by creating, using and managing appropriate technological processes and resources. The term educational technology is often associated with, and encompasses, instructional theory and learning theory. While instructional technology is "the theory and practice of design, development, utilisation, management, and evaluation of processes and resources for learning," according to the Association for Educational Communications and Technology (AECT) Definitions and Terminology Committee, educational technology includes other systems used in the process of developing human capability. Educational technology includes, but is not limited to, software, hardware, as well as Internet applications, such as wikis and blogs, and activities. But there is still debate on what these terms mean.

Technology of education is most simply and comfortably defined as an array of tools that might prove helpful in advancing student learning and may be measured in how and why individuals behave. Educational Technology relies on a broad definition of the word "technology." Technology can refer to material objects of use to humanity, such as machines or hardware, but it can also encompass broader themes, including systems, methods of organisation, and techniques.

Some modern tools include but are not limited to overhead projectors, laptop computers, and calculators. Newer tools such as "smartphones" and games (both online and offline) are beginning to draw serious attention for their learning potential. Media psychology is the field of study that applies theories in human behaviour to educational technology.

The word technology for the sister fields of Educational and Human Performance Technology means "applied science." In other words, any valid and reliable process or procedure that is derived from basic research using the "scientific method" is considered a "technology." Educational or Human Performance Technology may be based purely on algorithmic or heuristic processes, but neither necessarily implies physical technology. The word technology comes from the Greek "techne" which means craft or art. Another word, "technique," with the same origin, also may be used when considering the field Educational Technology. So Educational Technology may be extended to include the techniques of the educator.

A classic example of an Educational Psychology text is Bloom's 1956, *Taxonomy of Educational Objectives*. Bloom's Taxonomy is helpful when designing learning activities to keep in mind what is expected of—and what are the learning goals for—learners. However, Bloom's work does not explicitly deal with educational technology *per se* and is more concerned with pedagogical strategies.

An Educational Technologist is someone who transforms basic educational and psychological research into an evidence-based applied science (or a technology) of learning or instruction. Educational Technologists typically have a graduate degree (Master's, Doctorate, Ph.D., or D.Phil.) in a field related to educational psychology, educational media, experimental psychology, cognitive psychology or, more purely, in the fields of Educational, Instructional or Human Performance Technology or Instructional Systems Design. But few of those theorists would ever use the term "educational technologist" as a term to describe themselves, preferring terms such as "educator."

The transformation of educational technology from a cottage industry to a profession is discussed by Shurville, Browne, and Whitaker.

FORMS OF EDUCATIONAL TECHNOLOGY

In the past, learning and education simply meant face-to-face lectures, reading books or printed handouts, taking notes and completing assignments generally in the form of answering questions or writing essays. In short; education, learning and teaching were considered impossible without a teacher, books and chalkboards. Today, education and training have taken on a whole new meaning. Computers are an essential part of every classroom and teachers are using DVDs, CD-ROMs and videos to show students how things work and operate. Students can interact with the subject matters through the use of such web based tools and CD-ROMs. Moreover, each student can progress at his/her own pace.

How Important is Technology in Education

The role of technology in the field of education is four-fold: it is included as a part of the curriculum, as an instructional delivery system, as a means of aiding instructions and also as a tool to enhance the entire learning process. Thanks to technology; education has gone from passive and reactive to interactive and aggressive.

Education is essential in corporate and academic settings. In the former, education or training is used to help workers do things differently than they did before. In the latter; education is geared towards creating curiosity in the minds of students. In either case, the use of technology can help students understand and retain concepts better.

Factors that Help Students Learn Better

Research has shown, time and again, that students learn best when they are engaged. Through the use of technology, students can become active participants as opposed to passive ones where they simply receive instructions or information. Trust is another factor that enhances the learning ability of students. With the help of technology, teachers can establish credibility in what they are teaching. Web based tools can be used for providing demonstrations and examples that can help students establish credence in what they are learning.

Technology Allows Distance Learning

Perhaps the greatest impact of technology in the field of learning is its ability to help several people learn simultaneously from different locations. Learners are not required to gather at a predetermined time or place in order to learn and receive instructions and information. All one needs is a computer connected to a modem (or with a CD drive); these tools can literally deliver a 'classroom' in the homes and offices of people.

Technology Allows Group Learning

There are naysayers who argue that distance learning of this sort cannot help students receive the support of traditional group-based learning. For proving this theory wrong, technology has helped provide distance learners with online communities, live chat rooms and bulletin boards. All these allow students to collaborate and communicate even though they are isolated in their own space.

Technology Allows Individual Pacing

Multimedia tools, on-line and CD-ROM based training have helped eliminate the need for an instructor-based lesson plans. Students who grasp concepts faster proceed and move along, without being held back by ones who need more time and help for learning. Such individual pacing is beneficial to all.

Technology Helps Lower Training Costs and Increases Productivity

Another benefit of using technology to reach many students in shorter time is lowering training costs. Corporate and academic Institutions can reduce their costs of delivering lessons to students on a per-student basis. Moreover, technology produces quantifiable results and allows students to put into practice this information quickly and with better results. Through the use of technology, students can considerably save time and increase their productivity. Both these points justify the higher costs of advanced technological tools.

Roadblocks in the Use of Technology in Learning

Naturally, for education technology to have a positive impact on students, it should be designed and prepared well. Tools used for disseminating information must be developed with students in mind.

There are also factors like lack of computer/technology literacy to be considered. Schools and businesses must bear in mind that education technology is simply a tool and its success depends largely on the amount of planning that goes into it.

Using education technology can be a right choice as long as all such factors are considered.

TEACHING TEACHNOLOGY

DE serves as an alternative method for delivering academic coursework to students unable to attend traditional campus-based classes.

Mielke's discussion includes the following:

- *Definition*: DE is a method of education in which the learner is physically separated from the teacher and the institution sponsoring the instruction. Originally, DE involved teachers traveling to remote sites to teach classes or teachers corresponding with students. Technology has raised the quality of individualized distance instruction.
- *Forms of DE*: There are video and audio models of DE that involve broadcast and cable television, satellites, microwaves, fibre optics, and audio graphics. The linking of computer technology via the Internet or CD-ROM with television transmission provides a new dimension to DE, and can link professors to students in a distance setting. Another form of interaction is the use of computer conferencing. This method utilizes asynchronous communication in such forms as an e-mail list group, an Internet discussion group, or other types of conferencing software.
- *Adaptability*: Traditional programs that are heavily based in skill development and demonstration or require laboratory work can be offered in a DE format using interactive video interfaced with computers to facilitate a hands-on learning approach. Classes that use lectures and laboratory experiences are easily adapted to a DE situation.
- *Effective teaching and learning with DE*: DE dictates changes in behaviour for both the teacher and the learner; successful students develop

persistence and skills in self-directing work. Critical elements for successful distance teaching are:

- Instructor enthusiasm (comfort in front of the camera or with the technology);
- Organization (teaching materials must be prepared in advance);
- Strong commitment to student interaction;
- Familiarity with the technology used in the class format; and
- Critical support personnel (production staff, graphic designers, and technical staff members).

Educational Principles and Technology

Driscoll discuses four principles that offer a framework to teachers for thinking about how technology can support teaching:

1. Learning *occurs* in context, including the ways technology can facilitate learning by providing real world contexts that engage learners in solving complex problems, and computer simulations and computer-based micro worlds that offer contexts for learners to explore and understand complex phenomena in a variety of subject areas;
2. Learning is *active*, including the use of brainstorming, concept mapping, or visualization software, as well as simulations that enable learners to experiment with modeling complex ideas;
3. Learning is *social*, including software that supports a networked, multimedia environment in which students collaborate on learning activities; and
4. Learning is *reflective*, including technologies that promote communication within and outside the classroom, making it easier for feedback, reflection, and revision to occur.

Multicultural Education and Technology

In recent years, multicultural education and technology have emerged as key issues in education. Marshall describes how technology can support multicultural education efforts based on the five critical dimensions of multicultural education:

1. Content integration is intended to expand the curriculum by incorporating contributions of diverse cultures into traditional disciplines of study;
2. Knowledge construction promotes critical literacy by making explicit the manners in which scholars contribute to their respective fields of study;
3. Prejudice reduction is about eliminating all forms of bigotry and

involves promoting healthy personal identity devoid of the tendency to need to denigrate people who are different;

4. Equity pedagogy is about equalizing opportunities to learn, involving incorporating various strategies that attend to learning styles and intelligence types; and
5. DE (*i.e.,* everyone has access to specific courses or programs) is one example of empowering school cultures and social structures.

Therefore, teachers at all levels accept that technology has become an integral aspect of the teaching-learning process, but it is too soon to crown the duo a perfect pair.

TECHNICAL TEACHERS TRAINING

Even though often less acknowledged, the importance of a good teacher is enormous in India. They play a vital role in the overall development of the students. Not only are they responsible for imparting academic knowledge, but are also responsible for inculcating the right values and principles to their students. The importance of teachers is especially enormous during the formative years of children when they first join school. Therefore it is very important to have professionally qualified teachers to ensure the right development of students. At present there are several colleges and institutes offering Teacher's Training Courses in India. Since the needs of the primary students are different from the secondary students, the primary teachers and secondary teachers are required to take up different teacher's training courses. Therefore there are different teacher's training institutes offering:

- Basic training certificate (BTC)
- Junior basic training (JBT)
- Nursery Teacher's Training (NTT)
- Diploma in Education (D.Ed),
- Primary teachers training (PTT),
- Bachelor in Education (B.Ed)
- Other teachers's training courses.

Basic Training Certificate (BTC)

The quality of the education system plays a major role in the development of the country as it builds up the human resource of the future. But the sustenance of a vibrant educational system depends to a large extent on the quality of teaching at the elementary level. In order to improve the quality of the teachers, the government of India has set up various institutes that provide Basic Training Certificate (BTC) after the successful completion of a course in teacher's training. The institutes providing Basic Training Certificate

(BTC) courses are set up in almost all the states in India so that maximum number of teachers can avail the training and help improve the quality of education in India. The basic training certificate courses in the field of education are increasingly becoming popular as recruitment boards of primary teachers in various Indian states regard the BTC as a part of the minimum eligibility criteria for appointment of primary teachers.

There are several institutes that offer Basic Training Certificate (BTC) Courses in India for teaching at the elementary stage. On completion of the Basic Training Certificate (BTC) Courses from an institute in India, one can apply to various primary schools.

Junior Basic Training in Education

The concept of JBT-Junior Basic training in India is a result of the realization that education received at the elementary level shapes the human personality in more ways than one. It is therefore in the interest of the students that they be taught by trained teachers who can develop the cognitive skills and also take care of the emotional needs of the students. The junior basic training Institutes (JBT) have come up in India mainly to train the teachers who are actively engaged in imparting education to the tender aged children.

Eligibility and Admission Criteria for JBT-Junior Basic Training in India

In some of the states in India the JBT or Junior Basic Training course is conducted for a period of two years. Entrance test is also conducted for the admission of students to this course. Apart from the government institutes there are quite a number of private institutes that conduct junior basic training courses in India. A certain number of seats are also kept reserved as per the rules and regulations mentioned by the government.

Students who have passed their 12th standard board examination or equivalent are eligible for taking admission in this course. In some of the states in India a minimum percentage of 50% marks in 10+2 examination is essential for the students willing to join this course. Candidates with a bachelor's degree seeking admission to this course should have a minimum percentage of 45% marks. Candidates should be between seventeen to twenty eight years of age. Five years maximum relaxation in the upper age limit is given to the reserved category candidates.

Junior Teacher's Training Certificate Courses

As schools bring together students from different sections of the society, training of teachers is very important so that they can take care of the differential needs of the students. Junior teacher's training certificate courses are designed to impart training to the teachers so that they can make the students confident enough to face the challenges of the future. Teachers trained through JTTC are expected to be actively engaged in the whole process of producing enlightened citizens of the country.

Elementary Teacher Education (ETE)

Elementary Education refers to the education imparted to children between the age group of 4 to 14 years. The education system in India is supposed to be the second largest and well-developed after America but in the rural areas, the number of drop-outs at the elementary level is quite high. The government, however, is taking steps to enhance the scenario and has been creating programs that would make education easily accessible to everyone and also ensure that they continue with higher education. In order to strengthen elementary education in India it is also essential to provide proper training to the teachers entrusted with the duty of educating the students. There are quite a number of institutes that conduct elementary teacher education (ETE) in India.

Elementary level education lays the foundation of learning in a child. It is, thus, very important that teachers acquire proper training in order to handle children at the elementary level. The course includes all the major aspects of the field of study.

Apart from the theory part, teachers also go through practical training courses. However, candidates willing to pursue an elementary teacher education course in India need to fulfil the admission criteria as mentioned by the respective institutes. Candidates who have qualified their 10+2 examination or equivalent from a recognized board with the minimum percentage of marks are eligible for admission to this course. The growing need for trained teachers for the proper functioning of the schools have compelled private institutions to start various courses in elementary teacher education in India.

Primary Teacher's Training (PTT)

Well-trained teachers are capable of improving not only the quality of education in India but also the quality of life as they are actively engaged in building the human resource of the future. Primary Teacher's training is even more important as elementary education plays a very important role in a person's life. As they need to take care of the emotional needs along with developing the cognitive skills of the students primary Teacher's training is regarded as essential. Because of theses reasons Primary Teacher's training (PTT) is made compulsory in various states for appointment as primary teachers.

On this page we have tried to cover the institutes in India that offer the Primary Teacher's Training (PTT) courses.

We have tried to provide as accurate information as possible while compiling this list of PTT institutes in India. However, if you come across any discrepancy or error, do write to us about it. We would welcome any kind of feedback that will improve the quality of this page-a page that concentrates on providing the best information on PTT institutes in India.

Diploma in Education

As elementary education plays a very important role in a person's life and also in the development of the nation, the need to train the teachers is increasingly been realized. In India there are different institutes that provide Diploma in Education, D.Ed. This diploma in education would help the teachers in developing their cognitive skills and also in taking care of the psychological needs of the students. The D.Ed-Diploma in Education, India is gaining popularity as various states have made it compulsory to have D.Ed to be eligible for appointment as teachers. Many institutions offering D.Ed-Diploma in Education in India Teacher training in India is offered through a number of institutions across the country. The various institutions offering D.Ed-Diploma in Education, India include the following:

- Adarsh Junior College of Education, Maharashtra
- Adhyapak Vidyalaya, Junior College of Education, Mumbai
- Basic Training Institute, Raipur, Madhya Pradesh
- District Institute of Education & Training, Ambala, Haryana
- Hindu College Of Education, Haryana
- Padmashree Academy for Creative Teaching, Bangalore.

Qualification: For admissions to the D.Ed-Diploma in Education at various institutes in India, the minimum qualification required is completion of 12 years of basic education. Also 50% is the minimum marks required in the qualifying examination to be eligible for Diploma in Education.

Bachelor of Education, B.Ed

Teachers play a very important role in a student's life. It is, to a great extent, the teachers who decide the shape a student's life will take. So, it is very necessary to be adequately equipped with resources that will make the teacher a perfect role model to the students. To achieve this, Bachelor of Education or B. Ed was introduced, which will teach a person about teaching and the various aspects associated with teaching.

Once a person completes the Bachelor of Education (B. Ed) coaching or training, then he is awarded with a B. Ed degree.

There are some important reasons as to why one should opt for B. Ed course. After training one becomes efficient in teaching subjects of his specialization on the basis of accepted principles of learning and teaching. The course develops skills and widens understanding so that one can impart quality education to his students. The course also teaches about the attitude and makes him skilled in coming up with innovative teaching techniques. One becomes more competent in understanding psychological principles of growth and development and individual differences of the students he teaches.

There is a separate section in this course where the candidate learns to guide the children and counsel them in solving their personal and academic problems.

Top B.Ed colleges in India: If anyone is willing to have a strong foundation of teaching, it is always prudent to seek admission in some of the reputed B. Ed colleges in India.

Among them are:

- Jamia Millia Islamia, Maulana Mohammed Ali Jauhar Marg, New Delhi
- A.G. Teachers College, Ahmedabad
- Himachal Pradesh University, Department Of Education, Shimla
- D.M.College of Teacher Education, Imphal
- Andhra University, Visakhapatanam
- University of Bombay
- St. Xaviers College of Education, Patna
- College Of Teacher Education, Kozhikode
- NSS Training College, Pandalam.

Indian Universities offering B.Ed course Correspondence courses: Besides the above mentioned colleges, there are some universities which offer integrated course on Bachelor of Education.

- Annamalai University
- Bangalore University
- Jamia Millia Islamia Institure of Post Graduate Studies and Research
- University of Kerala, Institute of Distance Education
- University of Madras
- Maharishi Dayananad University
- Mother Teresa Women's University
- Patna University
- SNDT Women's University.

The demand for qualified and trained teachers all over the country has made the students choose B. Ed to ensure a good teaching job.

M.Ed (Master of Education)

As the education sector is attracting adequate attention in the age of structural adjustment and economic reform, Master of education (M.Ed.) is increasingly becoming a preferred career course for many Indians. This has also helped in the growth of the M.Ed. Career Colleges in India. The M.Ed Courses in India are generally pursued after the successful completion of the Bachelor of Education (http://www.indiaedu.com/b-ed-colleges/index.html) (B.Ed.) courses. It is a Post Graduate degree course offered at most of the District Institute of Education & Training (DIET) centers across India. In India M.Ed (Master of Education) is mainly offered by institutions that are recognized by National Council of Teacher Education, New Delhi.

It is a government body in charge of improving and implementing teacher education programmes in the country. It is essential to pursue the M.Ed. courses if you are interested in pursing a teaching career. M.Ed. is considered an important course as it helps the students to learn the education systems and patterns more profoundly and comparatively. In every state in India there are several institutions that offer Master of Education courses. The eligibility criterion for admission to this course is that candidates must pass their Bachelor of Education B.ED from a reputed institution in India.

Technical Teachers' Training Institute (TTTI)

Technical Teachers' Training Institute, Chennai is a resource institute established by the Government of India for quality improvement of technical education in our country and in the Southern region in particular. Technical Teachers' Training Institute, Chennai is a model human resource development Institute for planning, designing, developing, organizing and evaluating quality training programmes, research studies and learning packages for technical and vocational education, industry and community. The Institute strives continuously and vigorously to further enhance its sensitivity to environmental changes and reach greater heights of excellence through active collaboration with national and international agencies on Projects and Programmes aimed at quality improvement of technical education systems.

About TTTI

The Technical Teachers' Training Institute (TTTI), Chennai, was established in 1964 by the Government of India as a key catalyst institution for ensuring quality in technician education in South India comprising the states of Andhra Pradesh, Karnataka, Kerala and Tamilnadu and the Union territory of Pondicherry.

The mandate of the Institute is to take initiatives to offer need based HRD programmes through appropriate modes and develop curricula and instructional resources. It will also foster research and offer consultancy and extension services for the total development of polytechnics and other technical and vocational institutions, the Directorates managing these institutions, business, industry and service sectors and the community at large.

In carrying out these, the Institute will collaborate with national and international agencies interested in and/or deriving benefits from technical and vocational education including business, industry and service sectors. TTTI, Chennai an autonomous organisation has made substantial and significant contribution towards improving the quality of technical education in all its aspects. Though during the initial stages the emphasis was on training of teachers, over the years the emphasis has gradually changed to assisting the state governments and the polytechnics in the region towards improving their education process and products. This has led to diversification of the Institute's activities to suit the requirements of the clientele system.

EDUCATIONAL TECHNOLOGY AND KNOWLEDGE

Educational technology is the study and ethical practice of facilitating learning and improving performance by creating, using and managing appropriate technological processes and resources." The term educational technology is often associated with, and encompasses, instructional theory and learning theory. While instructional technology is "the theory and practice of design, development, utilization, management, and evaluation of processes and resources for learning," according to the Association for Educational Communications and Technology Definitions and Terminology Committee, educational technology includes other systems used in the process of developing human capability.

Educational technology includes, but is not limited to, software, hardware, as well as Internet applications, such as wiki's and blogs, and activities. But there is still debate on what these terms mean.

Technology of education is most simply and comfortably defined as an array of tools that might prove helpful in advancing student learning and may be measured in how and why individuals behave. Educational Technology relies on a broad definition of the word "technology." Technology can refer to material objects of use to humanity, such as machines or hardware, but it can also encompass broader themes, including systems, methods of organization, and techniques. Some modern tools include but are not limited to overhead projectors, laptop computers, and calculators. Newer tools such as "smartphones" and games are beginning to draw serious attention for their learning potential. Media psychology is the field of study that applies theories in human behaviour to educational technology.

The word technology for the sister fields of Educational and Human Performance Technology means "applied science." In other words, any valid and reliable process or procedure that is derived from basic research using the "scientific method" is considered a "technology." Educational or Human Performance Technology may be based purely on algorithmic or heuristic processes, but neither necessarily implies physical technology. The word technology comes from the Greek "techne" which means craft or art. Another word, "technique," with the same origin, also may be used when considering the field Educational Technology. So Educational Technology may be extended to include the techniques of the educator. A classic example of an Educational Psychology text is Bloom's 1956 book, *Taxonomy of Educational Objectives*. Bloom's Taxonomy is helpful when designing learning activities to keep in mind what is expected of—and what are the learning goals for—learners. However, Bloom's work does not explicitly deal with educational technology *per se* and is more concerned with pedagogical strategies.

An Educational Technologist is someone who transforms basic educational and psychological research into an evidence-based applied science of learning or instruction. Educational Technologists typically have a graduate degree in a field related to educational psychology, educational media, experimental psychology, cognitive psychology or, more purely, in the fields of Educational, Instructional or Human Performance Technology or Instructional Systems Design. But few of those listed below as theorists would ever use the term "educational technologist" as a term to describe themselves, preferring terms such as "educator."

The transformation of educational technology from a cottage industry to a profession is discussed by Shurville, Browne, and Whitaker.

2

Technology and Counseling Education

Emphasizing the fact that counselors use technology is largely unknown, Myers and Gibson report the results of the Association for Counselor Education and Supervision's survey on 12 technology competencies for counselor educators:

1. Using productivity software to develop Web pages;
2. Using such audiovisual equipment as video recorders, audio recorders, projection and playback units;
3. Able to subscribe, participate in, and sign off counseling-related Listserv;
4. Using counseling-related CD-ROM;
5. Using e-mail;
6. Using computerized statistical packages;
7. Using computerized testing;
8. Helping clients search for various types of counseling-related information through the Internet;
9. Having knowledge about the legal and ethical codes via the Internet;
10. Having knowledge about the strengths and weaknesses of counseling services provided via the Internet;
11. Using the Internet for continuing education opportunities in counseling; and
12 Evaluating the quality of Internet information.

Of the 92 counselor educators who responded, the three competencies with the highest levels were using e-mail, accessing Listserv, and using audio-visual equipment.

The three lowest competencies were using computerized testing, knowledge of Web counseling, and using computerized statistical packages.

SCOPE OF EDUCATIONAL TECHNOLOGY

By scope of educational technology we mean the jurisdiction, the limits or the boundaries within which it works. It needs demarcation of the boundaries within which the process of education can go on. Being a fast growing modern discipline it is almost practical all through and is expanding with a tremendous speed, aiming at all- round development in the area of education.

National Policy on Education (1986), recommends that, "Educational Technology will be employed in the spread of useful, information, the training and retraining of teachers, to improve quality, sharpen awareness of art and culture, inculcate abiding values, etc., both in the formal and non-formal sectors. Maximum use will be made of the available infrastructure."

The scope of Educational Technology can be accessed from the following points:

- *Determination of Objectives*: Educational Technology provides different methods and techniques for writing instructional objectives in behavioural terms such as Bloom Taxonomy Magar's Approach and RCEM Approach. The needs and requirements of the people and hence education need be revised from moment to moment. Educational technology helps in fixing-up the right objectives in the light of the changed circumstances and changed environment.
- *Improvement in Teaching Learning Process*: It helps in improving the teaching learning process and makes it more purposive. It tries to discuss the concept of teaching, analysis of teaching process, variables of teaching, phases and levels of teaching, principles of teaching, maxims of teaching and relationship between teaching and learning.
- *Development of Teaching Learning Material*: Teaching learning materials are also as important as anything else in the teaching learning process. In this age of science and technology, the materials of teaching cannot be unscientific. Everything of the society including values of life need be reflected in the materials. Only right type of material will be able to modify the behaviour of the learner suitably making him a fit person for the society.
- *Improvement in Teaching Training*: The change of environment with ne curriculum and new materials need be handled by the teachers. The teachers equipped with old strategies and methodologies of teaching will remain misfits. Right type of training to the teachers is the need of the hour. Educational Technology can render its valuable help in the training of teachers also. The use of video tapes and close circuit T.V. will help the teachers to remodel and reshape their teaching behaviours suitably. It includes micro teaching, simulated teaching,

term-teaching, teacher effectiveness, modification of teacher-behaviour, class-room interaction and interaction analysis, etc.

- *Development of Teaching Learning Strategies*: A strategy plays an important role in the hands of a teacher in every learning situation. The strategy has to be the right one which should be according to the materials and is able to bring about effective teaching- learning. The different strategies are being evolved by educational technology. The knowledge of those strategies is a must for every teacher. Then only the teacher will be able to do justice to their jobs. It tries to describe the ways and means of discovering selecting and developing suitable strategies and tactics of teaching in terms of optimum learning and available teaching-learning resources; the availability of the different types of teaching methods, devices and models of teaching-their appropriate selection and use for the optimum results.
- *Proper Use of Audio Visual Aids*: Audio Visual aids have always played an important role in the teaching-learning process. They need be used according to the times. The software aids, the hardware aids, the computer and other such appliances, equipment, etc., have to be used in the present type of teaching-learning environment. Computer assisted instructions will help the learner as well as the teacher to achieve the goals of education more conveniently.
- *Utilisation of the Sub-System of Education*: Educational Technology considers education as a system operating in a systematic and scientific way for the achievement of educational objectives. For the coverage of its systematic approach, it tries to include the topics dealing with the theory and principles of a system approach, explaining education as a system, its different sub-systems in terms of input and output. It is helpful in solving scientifically educational administrative problems with the help of system analysis.
- *Development of Curriculum*: Educational Technology is concerned with the designing of a suitable curriculum for the achievement of the desired objectives. It is helpful in describing the ways and means of the selection of suitable learning experiences, organisation of the contents in a suitable framework in order to bring better results. It provides the scientific foundation to education as well as develops theories of teaching and learning.
- *Proper Use of Hardware and Software*: These days' hardware and software are playing an effective role in the attainment of educational objectives. Educational Technology helps in the proper use of these aids. It tries to describe these resources in terms of their specific functions, their solution, proper handling and maintenance.

- *Provides Feedback*: It provides an appropriate feedback to the learners as well as teachers for bringing necessary improvement at the preparatory and implementation stages of their specific acts. For this purpose, educational technology discusses the ways and means of suitable evaluation techniques, their planning, development selection and appropriate use in relation to the objectives of teaching- learning system. In this way Educational Technology is concerned with all those who are connected in any way, directly or indirectly, to the processes and products of education. It teaches the teachers the art of teaching, the learners the science of teach the educational planners the structure of planning and administrators or managers the skill of managing or administering the task of teaching and learning. It works for the individualisation of instructions as well for improving the group dynamics of the class-room.

ELEMENTARY SCHOOL COUNSELLING

Elementary professional school counselors following best practices provide developmental school counselling curriculum lessons (Stone & Dahir, 2006) on academic, career, college readiness, and personal and social competencies, advising and academic/career/college readiness planning to all students, and individual and group counselling for some students and their families to meet the developmental needs of young children K-6 (ASCA, 2005).

Increased emphasis is starting to be placed on college readiness counselling at the elementary school level as more school counselling programmes move to evidence-based work with data and specific results (Dimmitt, Carey, & Hatch, 2007). Research has shown that school counselling programmes help to close achievement and opportunity gaps in terms of which students have access to school counselling programmes and early college readiness activities and which students do not (College Board, 2008). To facilitate the school counselling process, school counselors use a variety of theories and techniques including developmental, cognitive-behavioral, person-centered (Rogerian) listening and influencing skills, systemic, family, multicultural, narrative, and play therapy. Sink & Stroh (2003) released a research study showing the effectiveness of elementary school counselling programmes in Washington state.

Middle School Counselling

In middle school counselling, professional school counselors following best practices provide developmental school counselling curriculum lessons (Stone & Dahir, 2006) on academic, career, college readiness, and personal and social competencies, advising and academic/career/college readiness planning to all students and individual and group counselling for some students and their families to meet the developmental needs of late childhood and early adolescence according to sources such as the ASCA National Model (ASCA, 2005).

Increasing emphasis has been placed on college readiness counselling at the middle school level as more school counselling programmes move to evidenced-base work with data and specific results (Dimmitt, Carey, & Hatch, 2007) that show how school counselling programmes help to close achievement and opportunity gaps in terms of which students have access to school counselling programmes and early college readiness activities and which students do not (College Board, 2008).

Middle School College Readiness curricula have been developed by The College Board that can be used to assist students and their families in this process. To facilitate the school counselling process, school counselors use a variety of theories and techniques including developmental, cognitive-behavioral, person-centered (Rogerian) listening and influencing skills, systemic, family, multicultural, narrative, and play therapy. Transitional issues to ensure successful transitions to high school are a key area including career exploration and assessment with seventh and eighth grade students. Sink, Akos, Turnbull, & Mvududu released a study in 2008 confirming the effectiveness of middle school comprehensive school counseling programmes in Washington state.

High School Counselling

In high school, professional school counselors following best practices provide developmental school counselling curriculum lessons (Stone & Dahir, 2006) on academic, career, college readiness, and personal and social competencies, advising and academic/career/college readiness planning to all students and individual and group counselling for some students to meet the developmental needs of adolescents according to sources such as the ASCA National Model (ASCA, 2005).

Increasing emphasis is being placed on college readiness counselling at the early high school level as more school counselling programmes move to evidence-based work with data and specific results (Dimmitt, Carey, & Hatch, 2007) that show how school counselling programmes help to close achievement and opportunity gaps ensuring all students have access to school counselling programmes and early college readiness activities (Carey et al, 2008). High School College Readiness curricula have been developed by The College Board to assist this process.

To facilitate school counselling, school counselors use varied theories and techniques including developmental, cognitive-behavioral, person-centered

(Rogerian) listening and influencing skills, systemic, family, multicultural, narrative, and play therapy. Transitional issues to ensure successful transitions to college, other post-secondary educational options, and careers are a key area. The high school counsellor helps students and their families prepare for rigorous post-secondary education and/or training options (e.g. college, trade school) by engaging students and their families in finding accurate and meaningful information on entrance requirements, financial aid, recommendation letters, test-preparation and so forth. Professional School Counselors at the high school level spend much of their time helping students and their families monitor their progress toward graduation and being adequately prepared for post-secondary options including college. Some students now turn to private college admissions counselors specialized in college admissions but the ethics of so doing is open to great debate in terms of who has access to this funding and there is little research-based evidence of effectiveness on the part of these outside parties. The fees for these college admissions counselors can be as high as $30,000.

A framework for Professional School Counsellor responsibilities and roles is outlined in the ASCA (American School Counsellor Association) National Model (2005). Lapan, Gysbers, & Sun's (1997) study showed correlational evidence of the effectiveness of fully implemented school counselling programmes on high school students' academic success. Carey et al's 2008 study showed specific best practices from school counselors raising college-going rates within a strong college-going environment in multiple USA-based high schools with large numbers of students of nondominant cultural identities.

EDUCATION AND CERTIFICATION/CREDENTIALING OF SCHOOL COUNSELORS

The education of school counselors (school counsellors) around the world varies greatly based on the laws and cultures of specific countries and the historical influences of their respective educational and credentialing systems and professional identities related to who delivers academic, career, college readiness, and personal/social information, advising, curriculum, and counselling and related services..

In the United States, a professional School Counsellor is a certified educator with a master's degree in school counselling (usually housed in a Counsellor Education graduate programme) with specific school counselling graduate training including unique qualifications and skills to address all students' academic, career, college readiness and personal/social needs through the use of school counselling programmes that deliver specific measurable competencies. About half of all Counsellor Education programmes that offer school counselling are accredited by the Council on the Accreditation of Counselling and Related Educational Programmes (CACREP) and all but one are currently in the United States with one in Canada and one programme under accreditation review in Mexico as of 2008 and maintains a current list of accredited programmes and programmes in the accreditation process on their website. CACREP has identified in 2008 an interest in accrediting more programmes outside of the United States.

According to CACREP, an accredited school counselling programme offers specific coursework in Professional Identity and Ethics, Human Development, Counselling Theories, Group Work, Career Counselling, Multicultural/Diversity Counselling, Assessment, Research and Programme Evaluation, and Clinical Coursework in a 100-hour practicum under the supervision of both a school counselling faculty member and a certified school counsellor site supervisor (master's degree in school counselling or higher, and appropriate certification) and a 600-hour internship under the supervision of both a school counselling faculty member and a certified school counsellor site supervisor (master's degree in school counselling or higher, and appropriate certification) (CACREP, 2001).

CACREP released the revision of the Standards for 2009 in 2008, and made a major change moving toward performance-based accreditation including evidence of school counsellor candidate learning. In addition, in the 2009 standards, CACREP greatly tightened and enhanced the school counselling standards with specific evidence needed of how school counselling students receive education in foundations; counselling prevention and intervention; diversity and advocacy; assessment; research and evaluation; academic development; collaboration and consultation; and leadership in K-12 school counselling contexts. (CACREP, 2009).

Certification practices for school counselors vary around the world. School Counselors in the United States may opt for national certification through two different boards. The National Board for Professional Teaching Standards (NBPTS) requires a two-to-three year process of performance based assessment, and demonstrate (in writing) content knowledge in human growth/ development, diverse populations, school counselling programmes, theories, data, and change and collaboration. As of February, 2005, 30 states offer financial incentives for this certification.

Also based in the USA, The National Board for Certified Counselors (NBCC) requires passing the National Certified School Counsellor Examination (NCSC), which includes 40 multiple choice questions and seven simulated cases which assess school counselors' abilities to make critical decisions on the spot. Additionally, a master's degree and three years of supervised experience are required. NBPTS also requires three years of experience, however a master's degree is not required, but only state certification (41 of 50 require a master's degree). At least four states offer financial incentives for the NCSC certification (McLeod, 2005). Both certifications have benefits and costs that a school counsellor would want to consider for national certification. NBCC has credentials counselors in the United States and internationally.

TECHNIQUES OF COUNSELLING

The techniques used by the counselling should be according to the pupil's uniqueness and personality. Williamson has described the techniques of counselling under the following five headings—

Cultivating Self-understanding

The pupil or the client should possess the knowledge and understanding of his own abilities and responsibilities very clearly. The client should understand all these before using these abilities and responsibilities.

Therefore, the counsellor should have the experience of test-administration and interpretation of the test scores. The test scores provide a strong base for diagnosis and prognosis in the counselling process.

Advising and Planning a Programme of Action

The counsellor starts with the client's objectives and his attitudes etc. and indicates favourable and unfavourable facts or data. He weighs the evidences and he understands the fact why he is giving special suggestion to the pupil. Williamson believes that the counsellor should state his attitude with definiteness. He should not look like an indecisive. The counsellor does not fear while giving advice or direct suggestion because the pupil does not understand the utility of scores. Williamson has explained the following methods of advising the pupil after gathering the data—

(i) Direct Advising-In this, the counsellor gives his advice very frankly. This type of method is appropriate for tough-millded people who oppose any activity and who do not fear of failure.

(ii) Persuasive Method - This method is useful when the data indicate clearly some definite alternate. The counsellor analyses only the evidences and observes the alternate actions.

Establishing Rapport

When a client visits the counsellor for the first time, the counsellor should welcome his client. He should make him comfortable and he should be taken into confidence. The main basis of establishing rapport is-the fame of the counsellor's ability, respect for individuality, confidence before the interview and developing relations with the pupil.

Referral to Other Personnel Workers

No counsellor can solve all types of problems of the pupils. He should recognize his limitations and he should be given the knowledge of sources of specialized help. He should suggest the pupils to seek the assistance from other sources.

In addition to above mentioned techniques, there are other counselling techniques which are as follows—

Silence: Sometimes, silent-listening proves more effective in different situations. When a client is describing his problem, the counsellor adopts silence. This ensures the client about the counsellor's full attention towards the client's narration. It also ensures the client about the counsellor's serious thinking on the client's problem.

Acceptance: The counsellor should grant temporary acceptance to the client's case. Sometimes, the counsellor utters such words which give the impression that he understands very clearly whatever the client is stating. But the counsellor utters the words in such way that the client's talk remains uninterrupted for example, 'alright' 'very good', 'yes' etc. At many occasions, the counsellor does not utter any word for his acceptance! but he nods expressing his acceptance.

Clarification: At some occasions, the counsellor should clarify what the client says. It is the duty of the counsellor to make the client familiar with the fact that he understands the client and he accepts him. But sometimes it becomes necessary to clarify the client's explanation. While clarifying, the client should not feel any compulsion.

Restatement: Acceptance and restatement both ensure this client that the counsellor understands and accepts him. Through restatement, the counsellor repeats what the client has stated. But the Counsellor does not apply any correction or clarification during the client's assessment while restating, at his own.

Asking Questions: The counsellor should ask some questions to inspire the client to express more ideas. These questions should be asked just after the client's statement is over.

Approval: The client, expresses his different views regarding his problems. The counsellor approves some of them and disapproves others. The ideas which are approved affect the client very much. The client is also influenced by the knowledge possessed by the counsellor and his personality. If the counsellor 'grants approval to the client's ideas intermittently, then the approval becomes ineffective. Attention should be paid towards this.

Humour: Use of humour just to remove the client's tension and to make the conversation interesting has become the necessity of the day.

Summary Classification: Some part of the client's statement may not be useful Hence, the problem looks vague in itself In such situation, it becomes necessary for the counsellor to shorten the client's statement and to organize it properly so that the client may understand the problem more distinctly. The counsellor should always try not to add his own ideas.

Analysis: The counsellor can take initiative by presenting some solution to the client's problem. But the counsellor cannot force the client to act upon that solution. He leaves it upto the client whether to accept or reject that solution or to apply some modification. The client cannot be put under pressure in this regard.

Interpretations: The counsellor should have the right to interpret the client's statement or explanation only. He should not add from his own side. He concludes the result out of client's statement. The client alone is unable to derive results. It is important to note here that the results concluded by the counsellor and results derived" from the tests may agree or may not agree.

Regression: Sometimes it is erroneous whatever the client thinks or speaks. Such faulty ideas should be given up. The counsellor should act very carefully In order to give up such ideas so that the client may not turn agitated and the client may not think otherwise in negative sense.

Assurance: The technique of providing assurance to the client develops optimism in the client regarding the solution of his problem. Through assurance the counsellor approves the statements of the client and also provides support along with approval.

Effects of assurance are equally observed. Assurance is considered more comprehensive than the approval. Hence, the approval is included in the assurance.

Explanatory Method

Explanatory method is the most desirable method in the counselling. In this, the counsellor gradually understands very carefully the diagnostic data and indicates those situations in which the pupil's potentialities can be used. In this, the, use of data are explained in detail and carefully. Then after knowing the decision or interest of the client, the counsellor helps directly in order to execute the decision. This help includes the remedial work, educational or teaching plans etc.

EDUCATIONAL TECHNOLOGY AS SYSTEMS APPROACH

Educational Technology as a systems approach: All attempts made to define the concept of educational technology as an area of study involving the application of technologies emerged from the application of theories of learning and development as well as information and communication technologies have not been comprehensive enough without a theoretical grounding in the social context. The use of these technologies has to be grounded in a theoretical foundation provided by a systems perspective. The field of educational technology shares the same difficulties and struggles involved in defining itself as one comes across while defining other social sciences and applied social sciences. This section attempts to provide you with adequate theoretical understanding about systems theory in order to have a more comprehensive view of the field of educational technology.

You would have read through the two sections indicated in the article by Luppicini (2005). You would have understood that a comprehensive definition of educational technology goes beyond uses of technology including techniques, theories, and methods from multiple knowledge domains which are standardised and reproducible such as computer science, psychology and communications. The definition would also include the governing principles of systems approach.

'The systems approach to educational technology The systems approach to the design and analysis of teaching/learning situations is the basis of the great majority of modern educational technology-related developments. However, the terms system and systems approach are themselves jargon terms that can have a variety of interpretations. Let us therefore take a look at these terms in order to define the way in which we are to use them.

In general systems theory, a system is any collection of interrelated parts that together constitute a larger whole. These component parts, or elements of the system are intimately linked with one another, either directly or indirectly, and any change in one or more elements may affect the overall performance of the system, either beneficially or adversely.

The Systems Approach in Technology Education

The traditional approach in engineering or technology teaching is bottom-up, i.e. from component to system. For example, the order of the courses in a typical communications engineering program is: mathematics (calculus, etc.), science (physics, etc.), electricity basics, components, linear circuits, modules, basics of transmission and receiving, subsystems, and communications systems. In most traditional curricula, both in high school and undergraduate programs, the stage of dealing with a complete system is sometimes not fully addressed by the curriculum.

The larger, more complex, more dynamic and more interdisciplinary the specifications for a technological systems get, the harder it is for a lone engineer, as skilled as he or she may be, to design a complete system. Given this, students and their teachers, who are not required to be proficient in engineering, but who should be technologically literate, should not be expected to know so much as trained engineers as they go about manipulating entire technological systems.

Based on the systems thinking approach, what follows is a proposal for a way to teach technology and instill technological literacy without first teaching the details (for instance, electricity basics and linear circuits for electronics, or calculus and dynamics basics for mechanical engineering).

The central idea in this premise is that complete systems can be handled, conceptually and functionally, without needing to know their details. According to this approach, when trying to develop technological literacy in students who are not required to be proficient in engineering, the favored teaching strategy is top-down. In other words, the focus must be on the characteristics and functionality of whole systems and the interdependences of the subsystems.

System Approach

This new technology has influenced the educational administration and organisation to a great extent. This is the modern approach.

It acts as a link between hardware and software approach. It is also known as 'Management Technology'. It has brought to educational management a scientific approach for solving educational administrative problems.

It is essentially a new management approach, influencing management decisions in business, industry and education. Education is regard as a system and system approach is a systematic way of designing an effective and economical educational system.

System is defined in the dictionary as "an assemblage of objects united by some form of regular interaction or inter-dependence; as organic or organised whole as the solar system or a new telegraph system". System may be divided into three broad categories.

Systems Approach

System approach is a systematic attempt to coordinate all aspects of a problem towards specific objectives. Webster's dictionary defines a system as "a regularly interacting or independent group of items forming a unified whole." The characteristics of a system of may be explained with the help of an example – various parts of the digestive system may be called as components of digestive system. Every component of the digestive system contributes to as supports in functioning of the digestive system as a whole.

In the context of education, system is a unit as a whole incorporating all its aspects and parts, namely, pupils, teachers, curriculum, content and evaluation of instructional objectives. The teaching-learning process is viewed as communication and control taking place between the components of a system. In this case, the system is composed of a teacher, a student and a programme of instruction, all in a particular pattern of interaction.

The System Approach focuses first upon the learner and then course content, learning experiences and effective media and instructional strategies. Such a system incorporates within itself the capability of providing continuous self-correction and improvement. It is concerned with all elements of instruction including media, including hardware and software. Its purpose is to ensure that the components of the organic whole will be available with the proper characteristics at the proper time to contribute to the total system fulfilling the objectives.

In the systems approach to instruction, the teacher has to plan completely the utilization of selected resource material and the classroom activities. The teacher should have a good overall view of the subject, know his/her limitations, know all about his/her pupils and the individual differences in their learning capacities and plan accordingly. The system approach involves continuous evaluation of learning outcomes and utilization of knowledge gained by analysis of results of evaluation to suitably modify the plan of approach to achieve the stated objectives.

Major steps in the systems approach in education are: 1. Formulating of specific instructional objectives to be achieved and defining instructional goals, 2. Deciding appropriate media to achieve these goals, 3. Defining learner characteristics and requirements, 4. Selecting appropriate methods suitable for effective learning to take place, 5. Selecting appropriate learning experiences from available alternatives, 6. Selecting appropriate materials and tools required, 7. Assigning appropriate personal roles for teachers, students and supporting staff, 8. Implementing the programme, 9. Evaluating the outcome in terms of original objectives measured in student performance and 10. Revising to improve efficiency of the system to improve students' learning.

Advantages Of Systems Approach

i. Systems approach helps to identify the suitability of the resource material to achieve the specific goal.
ii. Technological advance could be used to provide integration of machines, media and people for attaining the defined goal.
iii. It helps to assess the resource needs, their sources and facilities in relation to quantities, time and other factors.
iv. It permits an orderly introduction of components demonstrated to be required for systems success in terms of student learning.
v. It avoids rigidity in plan of action as continuous evaluation affords desired beneficial changes to be made.

Limitations Of Systems Approach

i. Resistance to change. Old ways are difficult to erase. There is always resistance to any new method or approach.

ii. Involves hard work. Systems approach requires hard and continuous work on the part of school personnel. Some are not prepared for the extra load.

iii. Lack of understanding. Teachers and administrators are still not familiar with systems approach. Though it has been successfully implemented industry, it has still to make headway in education.

3

Technology in Special Education

The focus is upon the more sophisticated electronic hardware used for educational purposes. As might be expected, these developments are in the forefront of the field and, therefore, even the most advanced work is exploratory: their widespread use in schools is vet to be undertaken, annronriately so at this point given the developmental nature of the work. Yet, even the brief glimpse of the future afforded by the writings points to the possibility of a radical break with current practices in the education of exceptional children.

This section could well have been titled "Electronic Aids," since the devices to be discussed here only assist teachers in meeting instructional goals. As is well known, the aids have to be programmed in the light of clearly delineated objectives by persons who are well grounded in the learning characteristics of exceptional children and, as well, in conventional methods of presenting subject matter. Such requirements impose some burden on persons working with technological aids.

In view of these prerequisites, there appears to be little likelihood that the classroom teacher will be replaced—an apprehension sometimes expressed as a potential consequence of technology in education. Indeed, quite the contrary. A frequently stated advantage of this emerging technology, when properly researched and developed, is its "potential for giving the teacher time for the really important things, the things that cannot be done by a book or a machine" (Bright, 1967). It is clear that the teacher's role in the new technology will be altered: she is much more likely to be a diagnostician of learning problems and a specialist in the development and utilisation of multimedia instructional systems; she will be much more skilled in developing higher order conceptual and interpersonal skills and in dealing with certain unique and operationally defined curriculum programmes—objectives thought to be best dealt with by the live "flexible" teacher.

Of course, many teacher role adjustments can be expected to accompany the transition to the new technology. Problems of retraining, of skill obsolescence, and of fear of displacement are among the more obvious potential difficulties. The solution of these "human problems" may well be the most formidable obstacles to full implementation of instructional technology in the educational enterprise.

This section includes two articles. The first, by Blackhurst, describes applications of technology in a variety of areas of special education and in this writer's opinion, sets the tone for the special education of the future. Some specialists would take issue with Blackhurst's categorisation of disability groups in some instances. Nevertheless with greater clarity than any writing to date, his article highlights the prospects for technology in special education. The second article of the section, "A Vision of Technology and Education" by Oettinger, deals with broad problems of technology in education.

Blackhurst discusses the potential impact of technology on education of the visually handicapped, the cerebral palsied and other physically disabled, the deaf and hard of hearing, the mentally retarded, the mentally advanced, and children with behaviour disorders and learning disabilities. The critical concept in this article is the "student subject matter interface which describes any device that is used by the student in interaction with subject matter." Not infrequently the computer and other electronic aids represent the interface, although the author does not limit himself to these aids. Teaching machines, for example, treated as a separate section in the present volume, would represent an interface for Blackhurst. A number of the innovations suggested by Blackhurst have been implemented in experimental form and he discusses some of these in his article.

There have been several studies which represent applications of the student subject matter interface concept in special education including a project on audio-visual job training for the mentally retarded by Neuhaus (1964), an investigation of the use of the automated typewriter and the Wyckoff film tutor in teaching reading to the prelingually deaf by Schein (1963), and explorations in the use of the automated typewriter in teaching the disadvantaged by McSweeney, Richardson, and Gotkin (1967). Each of these studies discusses the use of relatively sophisticated electronic equipment in presenting academic or vocational subject matter to exceptional persons. Each of the devices requires some active subject participation and also provides or makes provision for immediate feedback on the correctness of the students' responses, offering some unique advantages.

Neuhaus has written about a programme designed to teach relatively advanced industrial skills to mentally retarded adolescents and adults. The training device is the Videosonic audio-visual aid which presents coloured slides in conjunction with audio-instruction for each slide. The worker performs each step of the industrial operation as he sees and listens to examples of the step to be performed. A unique feature of the device is its potential for engineering retardate attention and its provision for repetition of examples of the skills to be mastered. These are principles known to be effective in the education of mental retardates but ones which are often difficult to effect in the typical classroom setting.

The Neuhaus study and those of Schein and McSweeney et al. highlight a critical theme running through the entire volume, *e.g.*, we are witnessing in special education more careful attention to specific aspects of learning and behavioural processes in exceptional children. In the plug soldering programme described by Neuhaus, for example, 216 slides divided into six sections had to be developed "a very thorough process of job simplification." The experimenter was forced to look carefully at each step in the instructional process. These principles hold, of course, for all programmed instruction, for operant techniques where specific carefully delineated aspects of behaviour rather than broad ones are modified, and in the assessment of psycholinguistic functions, where specific psycholinguistic abilities rather than global ones are assessed. We are witnessing a movement, although all too slowly, towards a more miscro-scopic analysis of the instructional process. The work using the more sophisticated electronic devices, behaviour modification, the assessment of psycholinguistic abilities, as well as programmed instruction and teaching machines presage this trend.

A rather interesting electronic aid which has been used experimentally in the education of exceptional children is the computerised typewriter, sometimes referred to as the talking typewriter (ERE).

The Edison Responsive Environment typewriter consists of a standard-size keyboard with coloured keys, an exhibitor for display of printed matter (with pointer), a miniature screen, and dictation equipment. Only the keyboard is accessible to the child. Once the key is depressed, a large letter, number, or punctuation mark appears on the paper, and a voice names it through the loudspeaker.

A number of variations are possible. For example, the respondent may be required to respond with a specific number, letter, or sequence of numbers or letters. The typewriter responds only if the proper sequence is followed. Already, the responsive environment has been used to teach reading and related skills to the culturally disadvantaged, the educable mentally retarded the gifted, the speech impaired, the emotionally disturbed and those with learning disabilities. Moore (1963), who with colleagues is largely responsible for the ERE, outlines the system's features:

1. It permits the learner to explore freely.
2. It informs the learner immediately about the consequences of his actions.

3. It is self-pacing, *i.e.*, events happen within the environment at a rate determined by the learner.
4. It permits the learner to make full use of his capacity for discovering relations of various kinds.
5. Its structure is such that the learner is likely to make a series of interconnected discoveries about the physical, cultural, or social world.

McSweeney, Richardson, and Gotkin (1967) have presented the results of a project using the ERE with disadvantaged pre-school children. Their study goes well beyond the mere question of whether or not the computerised writer facilitates the acquisition of pre-reading skills in the pre-school disadvantaged (it does) to more basic questions of motivation and other problems related to use of such devices. This work highlights the point that we are yet in the exploratory stage in the use of technology for educational purposes, and that there remains the need for programmes of basic research concerning their effective utilisation.

A variation of the ERE is described by Schein (1963). In his project pre-school deaf children actually engage in a language programme using the automated typewriter and the Wyckoff film tutor (a teaching machine). In this investigation the basic format of the ERE remains, but some modification is needed to take account of the subjects' auditory limitations.

Dupress (1967) has surveyed a variety of promising applications of technology to problems of the visually limited. These include reading machines for the blind, computer methods for implementation of high speed duplication of braille, and a variety of mobility aids based upon sonic, ultrasonic, laser-like, infrared light, and related principles.

Carr (1964) has conducted a study on tele-teaching with elementary and secondary homebound pupils. Tele-teaching is an instructional arrangement whereby homebound pupils receive instruction in their homes via the telephone. The procedure has been in use for more than a decade, but recent developments in technology have made Tele-teaching even more feasible.

The studies by Neuhaus, by McSweeney, Richardson, and Gotkin, and by Schein are concerned with subject matter presentation in which students interact with an electronic device in a learning situation.

The Tele-teaching arrangement, on the other hand, only facilitates student-teacher contact among pupils in diverse physical locations. This is in itself an important development, even though this teaching arrangement is less than ideal. Carr has, however, clearly forecasted developments which may lead to a closer approximation to the ideal teacher-student relationship—the use of the electro-writer, and the new picture phone in conjunction with Tele-teaching.

In the other article of this section, Oettinger presents a view of the potentialities and limitations of technological advances in education. Notice that even in this article written for a general audience, the author looks at the computer's potentialities for individualizing instruction with students falling at the extremes of the ability continuum (gifted and mentally retarded) and with culturally disadvantaged students. What seems most noteworthy, however, is Oettinger's analysis of broad issues and problems with which technology in education must deal-problems of funding, of staffing, of the potential negative influence on other segments of the economy, and even of thought control.

His section "Cold Water" deserves special attention, as it might well apply to a number of topics treated in this volume. Technology is new to special education. At present most work in this area is exploratory. With increases in technological sophistication, and as special educators become more aware of its potential, we can expect that technology will play an increasingly important role in the education of exceptional children. However, as Suppes (1967), has noted:

> *As with any new technology with a potentiality for solving old problems, the initial development and implementation create nearly as many problems as are solved. The road ahead is bumpy and tortuous, but the signs point upward towards educational excellence. Many of those most concerned with the preservation of humanistic or individualistic traditions view the increasing importance of... technology in our society as a great threat to these traditions. If such critics understood better the scientific world in which we live, they would realise that this technology offers a possibility of individual fulfilment in education hardly conceivable fifty years ago. It is not too much to claim that for the first time since public education for everyone became a goal in our society, individualised instruction at a genuinely deep level is now a feasible goal.*

DEVELOPMENT OF SPECIAL EDUCATION – A GLOBAL PERSPECTIVE

The welfare of the handicapped has come a long way since the days of purely custodial and protective care. At first, it was believed that handicapped children were sub-normal and consequently, provisions were custodial in nature. But, eventually an attempt at rudimentary education and training became a part of the pattern.

Special education in western society had a true beginning in sixteenth century. Pedro Ponce de Leon taught some deaf pupils in Spain to speak, read a write. The first book on teaching the deaf was published in 1620 in Spain by Juan Pablo Bonet. This gave rise to a wider interest in the education of the deaf in Europe. In England, John Bulwer (1654) wrote on teaching the deaf to speak and read lips. In France, similar work was later carried out by Charles Michael, abbe L'Eppe (1912-89) who made profound contribution in developing the natural sign — Language of the deaf. His work was further developed by abbe Sicard (1742-1820) and gave rise to silent method of teaching the deaf. In Germany, Samuel Heinicke (1727-90) educated deaf children orally, and later Friedreich Moritz Hill (1805-74) developed this method which ultimately became accepted practice throughout the world.

No real attempt was made to educate and train the blind until the close of 18th century. Valentine Hauy (1745-1822) opened the first institution for blind in Paris in 1784. Hauy's success soon spread to other countries. Schools for the blind were opened in Liverpool (1791), London Vienna (1805), Berlin (1806), Amsterdam and Stockholm (1808), Zurich (1809) and Boston and New York (1832).

Scientific attempts to educate the mentally retarded began with the efforts of Jean-Marc-Gaspard Itard, a French Physician and otologist. Later, Edouard Seguin, a student of Itard who emigrated to US devised an educational method using physical and sensory activities to develop mental process. Sequin's published works influenced Maria activities to develop mental processes. Sequin's published works influenced Maria Montessori, an Italian pediatrician who became educator and innovator of a unique method of training young mentally retarded and culturally deprived children.

The field of special education for the handicapped children had in important beginning in the nineteenth century as represented in the pioneering work of Itard, Sequin, Gallandet, Hauy, Montessori, Braille and many others. In the United States many residential schools for deaf, blind and retarded children were established during the late nineteenth century. Early in the twentieth century, development regarding both the handicapped and gifted children were modest at best.

In the recent years, a remarkable upturn in the pace of development has been observed. Special education programmes in many countries have doubled and redoubled in size in the last two decades. In a summary of a series of nationwide surveys conducted were enrolled in special education programmes in 1948. The comparable figure for 1963 was, 1,666,000 which indicates that special education enrollments had more than tripled in 15 years. By 1963, about one-half of the school system in US was thought to be providing some type of special education service.

From the modest beginning in late nineteenth and early twentieth century, the field of special education, in recent years, had received greater attention due to the growing acceptance of the concept that a handicapped child is, in majority of cases, a normal child faced with some disadvantage which is often only temporary.

This new attitude is not yet universal and consequently the extent of activity at the national level varies from country to country. The International Conference on Public Education convened in Geneva by the UN's Educational, Scientific and Cultural Organization, and International Bureau of Education, held on July 6, 1960, justified organizing special education for handicapped children on the following grounds:

1. Right to Education laid down in the Universal Declaration of Human Rights applies to all persons capable of benefitting from it, including the least gifted.
2. Principle of compulsory education for all children is to be universally accepted.
3. It is an economic, social and cultural disadvantage for a nation to include a number of handicapped persons of whom might have become useful citizens if they had been suitably educated.
4. Any particular ability possessed by a handicapped child should be used to further his education.
5. Progress in medicine, child psychology and remedial education had made it possible to identify children suffering from various handicaps and to undertake their education by methods based on differentiated and individualized teaching.
6. The use of differentiated education method should not led to the complete social isolation of children who need to remain in contact with normal life so that they may later become integrated in it, and that the experienced teacher may take advantage of this situation to cultivate in children different abilities which contribute to greater mutual understanding.
7. The problem of existing special education for handicapped children is always related to the stage reached in providing education for normal children and to differences existing between rural and urban areas as well as to male and female population.

In advanced countries today, special education of the handicapped has become relatively universal. The developing countries are also deeply concerned about it and are showing ever increasing interest in the development of education and rehabilitation services for the disabled.

Voluntary associations both at the national and international levels are busy in contributing to the betterment of the lot of the handicapped. In order to promote general awareness and provide greater impetus to the special education and rehabilitation programmes of the handicapped at international level, the UN declared 1981 as the year of the disabled and 1983-92 as decade of the disabled.

GUIDELINES TO ASSIST TEACHERS OF MENTALLY HANDICAPPED CHILDREN

Guidelines were derived from the results to assist teachers of mentally handicapped children in coping with the challenges and needs as revealed by this investigation. Guidelines in respect of the *first theme*, special education is challenging by its very nature include the following:

To address the challenges posed by the different levels of development, intellectual abilities and languages of the learners in one class, the following strategies and approaches could be employed:

(i) The employment of additional staff: The employment of experts and assistants, such as psychologists, occupational, physio- and speech therapists, as well as nurses and class aides, whether on a permanent or temporary, paid or voluntary basis, would make a significant contribution to alleviating the pressure on teachers at special education schools.

(ii) The use of a variety of classroom strategies: The teacher could consider a variety of methods of presentation in the class. The volume of work to be taught could be reconsidered and the work could be arranged in smaller chunks (Engelbrecht and Green, 2001). Helpful devices, such as audio-tapes, video recordings and computers could also be incorporated. A well-organised teacher will probably also find it easier to handle the diversity in the classroom, eg early arrival to prepare the classroom and *to-do* lists of priorities.

(iii) Different groupings with appropriate curricula: Age-appropriate curricula are suggested to facilitate the differences between the phases. In the junior phase (6-9 years) the focus can be more on perceptual and motor, safety, communication and socialisation skills, body image and basic reading and numeracy skills (Engelbrecht and Green, 2001). In the middle phase (9-12 years) the focus can be on socialisation, self-sufficiency, communication, motor co-ordination and more functional academic skills, which allow greater independence. In the senior phase (12-18 years) career-related activities, such as handwork, painting and cooking can be introduced, as well as sexuality education.

The later years of the senior phase can be seen as a preparation for an occupation. Learners with profound mental impairment can be focused on practical skills, such as self-care, motor development, socialisation and safety. On-the-job training and part-time placements can be arranged for older learners (Hegarty, 1993)

(iv) An aappropriate record system: In order to enable teachers to place learners in the appropriate group, proper records of progress and achievement should be kept. A personal information file should be opened for each learner, as well as a personal assessment portfolio. This record should accompany the learner throughout his learning path and should be retained as visible proof of his development (Government Gazette, 1998).

(v) Appointment of tutors and class assistants: With regard to the language dilemma faced by teachers, they could appoint peer tutors as an invaluable source of assistance (Hallahan and Kauffman, 1997). In this way pupils could effectively tutor one another and develop peer networks. It is also an inexpensive source of assistance. Parents and other community volunteers could also stand in as class assistants for language translations (Williams and Snipper, 1990).

It was further evident from the results obtained in Theme 1 that discipline was a challenging issue for teachers in special education. Various ways and means of addressing the issue of discipline can include (Hallahan and Kauffman, 1997; Bos and Vaugn, 1994; Steenkamp and Steenkamp, 1992; Bernard, 1990): a code of conduct (disciplinary policy) for the school, classroom rules, routine and order, a balanced approach, flexibility, class mentors, setting up a reward system, stronger capacity, prefects or monitors, collaboration with families, and behaviour management workshops. The results obtained under Theme 1 also indicated that special education involved extra work and responsibility, and therefore causes stress. Various strategies and techniques could be employed to address this problem: Systematic planning is essential (Gulliford and Upton, 1992). The teacher also has to be resourceful, as she has to change an activity or instructional material midstream when it is clear that a planned programme does not have the desired results (Steenkamp and Steenkamp, 1992). Teachers should remain students to keep up with the latest developments in the field of special education. Teachers could also benefit from life skills training, by developing a positive outlook on life, and good interpersonal relationships (Dunham, 1992), to handle the stress they experience. To address the reality of the stigma attached to special education, improved interaction between mainstream and special education schools as well as the community and society at large, can be suggested.

Guidelines in respect of the *second theme,* Teachers in special education display special needs include orientation courses, adjustment, experience, specialized training and mentoring programs for teachers (Dunham, 1992).

1. To address the identified need for orientation, management could design and implement a preparation and adjustment program and mentoring for new teachers (Donald, Lazarus and Lolwana (1997) to ensure that the teacher assumes his/her duties in a productive manner, without insecurity.

2. Specialised teacher training by tertiary institutions and through in-service workshops is the appropriate avenue to equip teachers with the skills and competencies they need to address their task.
3. Effective in-service staff development workshops are vital for the development of appropriate teaching strategies. These workshops should be preceded by a needs assessment. The following, according to Engelbrecht and Green (2001), are some categories of activities that may emerge during staff development initiatives:
   - An analysis of barriers to learning and participation.
   - Deciding on a target issue.
   - Problem-solving and solution finding.
   - Setting of success criteria.
4. Strategies that could be adopted to provide teachers in special education with the required support are
   - teacher support groups (Dunham,1992)
   - a staff welfare policy (Bernard, 1990)
   - peer group support (Bernard, 1990)
   - voluntary expert support from the community.

An urgent requirement of teachers is active support from the Department of Education through the following:

- The Department could send delegations around to schools to advise and support teachers on relevant issues.
- Officials of the Department should give demonstration lessons in order for teachers to observe how to work with a mentally handicapped child.
- Workshops could be arranged to enlighten teachers on curriculum matters.
- Supporting means, such as finance, material and resources, are very important and should be provided by the Department of Education.

Programs for the development of parent support, empowerment and co-operation could be facilitated through home visits (to establish contact and interaction) and parent-teacher partnerships. Teachers could encourage parents to play a more active role at the school, such as painting the school or providing the school with legal advice, depending on the skills and competencies of the parent (Engelbrecht and Green, 2001; Shea and Bauer, 1985).

From the results it would seem as if there is a serious need for support from the community. Community support would include family members, volunteers, non-governmental organizations and the business sector. The following are possible activities and strategies to obtain community support:

- Awareness and information campaigns. Newsletters and brochures could be sent to businesses, firms and churches.
- Media campaigns. Articles could be placed in the local community newspapers on the achievements of and challenges facing mentally handicapped children.
- Involving the community in school activities, eg by means of invitations to open days and involving the learners in community projects, eg making greeting cards for old age homes.
- Using the school as center for community development, such as education programs for illiterate members of society, and making the school facilities available for community use, eg the hall, the sport fields (Donald, Lazarus and Lolwana, 1997).
- Utilizing community services. The community could share their expertise by offering art, dance or music lessons to the mentally handicapped child and exposing him to social experiences through concerts and performances in the community.

5. The need for resources in the school for the mentally handicapped child has to be addressed to prevent the child from suffering. This could be done by means of a resource network and sponsoring bodies (Norwich, 1990:37), as well as twinning relationships with other schools.
6. The need for strong leadership could be fulfilled by the principal, through understanding, support, good relationships, involvement of teachers and a democratic management style. The principal can consider

- discussing the vision for the school with the teachers;
- the educational goals to be striven for;
- a school leader development program;
- good communication channels (regular meetings, democratic decision-making and participation of teachers)

PROMOTING EDUCATION AND TRAINING OF DISABLED PEOPLE

UNESCO's Global Monitoring Report on Education for All (EFA) published in 2007 estimated that 77 million children (aged 6 – 11 years) do not attend school and that approximately one-third of these out-of-school children are disabled. The other two thirds are said to be children from poor families who live in poor households and whose mothers have no education. The above assertion follows up the World Summit for Children in 1990 and the adoption of the UN Convention on the rights of the child which states that more than 155 countries have developed National Programmes of Action. But despite all this international activity, few of the necessary changes in promoting education and training of disabled people have been made. Of importance to note has been the failure to address the barriers that make it difficulty for children to access education and training easily. Policies enacted by Governments remained on paper and were never implemented. Despite the concern that education should be accorded to all – still that continued to be ignored.

If ever good education is to be promoted, there should be deliberate foci on relevant policies that concern themselves with "right to access an inclusive, quality, free primary and secondary education on an equal basis with others in the communities in which they live" (UN Convention on Human Rights Article 24).

Problem Statement

Both Governments of Lesotho and South Africa have favourably responded to the United Nations Conventions on need for; observance of children's rights and provision for Education for All (EFA) inclusive of integration and mainstreaming of children with disabilities. In the two countries there are white paper Policy Statements which cater for placement of children with disabilities into schools. The good intentions illustrated in these white papers are never or partially implemented. One parent of the disabled in KwaZulu Natal said; *"Good ideas from these white papers remain in offices and never taken down to the grassroots – Us parents as implementers"*. This research project, therefore, will aim to conscientise the policy makers on the gap between what they say and what should be done on the ground.

Purpose of This Study

A number of studies, complemented by interesting evidence gathered during the research, demonstrated that disabled people are less likely to complete basic primary education than their non-disabled counterparts. This lack of basic education reinforces their social exclusion and marginalization thus leading to lack of life requisite skills making it more difficult for them to secure long term sustainable employment. This study therefore tried to find out whether basic education was accessible to disabled people and whether that basic education was adequate to help disabled people get training in the learnership/vocational programmes offered in South Africa and Lesotho. Respondent anecdotal evidence indicated that basic primary education without high school education was inadequate and insufficient to help anybody enter into vocational and learnership programmes – so only a small number of disabled children access training to enhance their skills and independent living. This dire dearth is there despite the good white papers produced by both Governments – South Africa and Lesotho that; For example, the South

African white paper on Integrated National Disability Strategy with a vision commitment of a *"better life for all persons with disabilities through excellence in governing"* and a mission which *"ensures the equalization of opportunities for persons with disabilities in leading, developing and managing the strategic agenda of Government"*. All these good ideas are mere pronouncements without implementation.

METHODOLOGY

Design

The qualitative and situational analysis type of design was used – with an integrated participatory approach. This approach is a generic research approach in which the point of departure is the *"insider perspective"* on social action so that attempts are made to study human actions from the *perspective of social actors themselves* (Monton 2001:125). This participatory approach used brought a stance of openness and discovery. Also this participatory approach is about, *"stories from the field"* and allows novices and scholars to see how critical, interpretive research is actually conducted. This approach used in this research is in fact reflective, respecting the openness to other views and perspectives and did not only include explanatory hypothesis but other people's ways of thinking and valuing.

The employed qualitative design in this investigation assisted the researchers to study the reactions of respondents – physically disabled, the blind, the deaf and the slow learners (mentally retarded). The use of in depth interviews helped researchers in two ways; a). to exchange real experiences with the different categories of disabilities. B). to live the lives of disabled people *"as they narrated their live, real stories"*. The researchers co-participated in the lives of the respondents; that is learning *"life by life"* (Elphick 2002).

Population

In both South Africa and Lesotho over 250 respondents formed part of the population of study. The population could have been bigger as planned but unfortunately after working on the ground the quantitative instrument (the questionnaire) could not be used due to illiteracy. Researchers after visiting two institutions found the respondents not literate enough to understand the questionnaire so the researchers quickly switched on to the interviews which worked well.

In sampling, the researchers purposely chose learning institutions – three in Gauteng, one in Mpumalanga, one in KwaZulu Natal and one in Lesotho. Two of those institutions were vocationally oriented.

Descriptions of the Sample (Population)

The researchers deliberately chose the different disability categories : Of the three institutions in Gauteng – one was a centre for slow learners, one was a rehabilitation centre whereas the other one was a Cheshire Home. The Mpumalanga Institution was a sheltered Workshop but mostly community run – also with a bigger population of respondents from age 11 to almost 65. In Maseru – Lesotho, the institution visited was a Catholic rehab centre with all types of disabilities. The KwaZulu Natal population was quite big and consisted of parents of disabled children. All categories of disabilities were there.

Almost 65% of the respondents in both countries were disabled young men and women within the age ranges of 18 and 25. 15% were disabled adults, 10% were parents of disabled people, 5% were teachers and those who directly helped the disabled people and 5% were DPOs representatives including DPSA president and LINFOD president in Maseru – Lesotho who at the same time is the Chairman of SAFOD regionally. It is clear that the sample did not include the education officials due to political gimmicks which may have influenced our sampling. But sadly some officials bulldozed their way in and we had fruitful interviews in Mpumalanga.

FACULTY RELUCTANCE WITH TECHNOLOGY INTEGRATION

The Problems Defined

Faculty reluctance, it appears, is often related to actual fear of technology, in addition to the research that suggests that the promotion and tenure review process often does not consider technical contributions as scholarship. Technical pockets of reluctance can be enabled by institutional attitudes, workload allocations, the time required to develop technical initiatives, and the promotion and tenure review climates. The responsibility for change rests with the entire set of stakeholders and classrooms where ultimately no hiding place from technology will remain, and students who will no longer tolerate a technological mismatch within the teaching and learning environment. IT has made swift changes in the past thirty years that have made a dramatic and noticeable impact on business, industry, and education through the use of computers and related technologies.

Employment opportunities in the next century will, for the most part, demand some computer application literacy. Applicants without an acceptable level of computer literacy will be disadvantaged.

Institutions of higher education and the faculty workforce associated have not been without impact of this rapid and dramatic change. With the proliferation of the new technologies, many professors are undertaking what they perceive to be innovations in their teaching by incorporating differing uses of technology into their instruction. Technology and the Internet force universities to change how they deal with the growing technical needs of these organizations. This is particularly evident now, as students entering college may be better equipped technologically than are their faculty. Beyond budgetary and infrastructure deficiencies relating to digital information access, an additional obstacle exists in terms of maximizing technical academic capabilities.

Too often pockets of technically reluctant faculty attempt to hide from the ground swell of the digital current that is engulfing higher education. By dragging their feet for a variety of reasons, it has become apparent that the ripple effect of their reluctance impedes and conflicts with the direction that information access is taking today. Within the context of DE and the academic cultural environment of higher education, technology appears to have altered professional relationships.

Traditionally, institutions of higher learning have been authority- and knowledge- centered: academia is a controlled hierarchical system for delivering knowledge from authorities to students. The provision of DE foreshadows a fundamental shift to a more lateral structure. Innovative and far-reaching views on the proposed development and integration of instructional use of technology relating to learning and teaching are repeatedly featured throughout the profession. In preparation for this technical transformation, the Faculty of Education of the University of Macau, conducted research relating to programme concepts and skills required to develop and sustain instructional technology delivery in China. The FED has offered a declaration in the form of a key mission to be realised in the future years, involving all students and staff, that all teachers preparing for the information age should be conversant with knowledge and applications of IT in the classroom teaching and learning. According to Kwok-cheung, by taking the lead in developing instructional technology, the purposes of the Macau University technology initiative were:

- To establish an affective understanding of computers;
- To understand applications of computers in daily lives and the importance of computers in modern society; and
- To encourage interests and consciousness in the use of computers.

Therefore, the FED concluded the immediate necessity to develop texts and teachers' guides to support these instructional initiatives.

An Investigation of Faculty Technology Skills

Lamboy and Bucker describe some interesting findings related to faculty computer skills in their paper *An Investigation of Faculty Technology Skills in a Puerto Rican University.* The purpose of this study was to evaluate full-time faculty technical expertise and computer competence in order to approach the implementation of Web-based opportunities for instruction. This private nonprofit university offers academic programs in liberal arts, education, business administration, science and tourism, leading to baccalaureate and associate degrees, as well as masters' degrees in human resources, administration and supervision in education, marketing, criminal justice and management. The student body includes recent high school graduates as well as adults who are returning to school for professional development.

This university recently completed a licensing review and was recognized favourably in three areas:

1. Professors, when exploring new initiatives, enjoy support for the Chancellor;
2. As the administration supports new approaches and techniques, the faculty is exploring new methodologies for instruction; and
3. An exceptionally broad range of services are available for the student body, including professional counseling, tutoring, educational services, and mentoring.

Further, a preliminary investigation on the University study relating to attitudes on faculty technical competence provides valid points for review, such as:

- Technology is evasive for changing the process of education and the role of the faculty.
- DE has become a point of discussion with the majority of colleges and universities.
- The issue of quality and its distinction between traditional on-campus and DE is a concern, as is addressing cost and student diversity? Administrators have high expectations from investments in technology and Internet services.
- Higher education IT initiatives suggest that many institutions provide Web access undergraduate catalogs and applications as well as library-based reserve materials, in addition to courses offered via the Internet.
- The Internet is a viable partner with instructional methods by accommodating the academic profiles of a diverse student population.
- Educational technology is forcing the traditional teaching approach towards addressing new skills that are becoming essential for both faculty and student survival.
- Maintaining a current level of technical competence is often a major source of stress for faculty.

The Internet and future visions of education technology applications are applauded and promoted to enhance the delivery of teaching methods and to facilitate streamline routine instructional preparation and administrative tasks. This advancement arouses concerns. "A substantial number of an institution's faculty members must have an understanding of and some experience with integrating elements of a computer-based instruction into existing courses".

Twigg continues by stating that institutions and faculty may have enthusiasm for initiating new technologies, but to justify the investment and to maintain quality of instruction the IT expertise must be in place.

In addition to promoting new technologies and providing appropriate expertise, institutions must also allow for and value the time required to develop and deliver technical instructional applications. Rajagopal and Bojin discuss the establishment of the Institute for Learning Technologies provided by Robert McClintock, which indicates that the integrations of IT often have a variety of outcomes.

It appears that academicians, often expected to incorporate IT consistently while carrying heavy workloads, are challenged to find the time to master and implement new technologies.

Research, teaching, and service are the mantra for achieving promotion and tenure, but faculty at most traditional U.S. universities find that research is critical while other factors play a supporting role". Institutions of higher education face a challenging debate involving pressure to allocate funding for the development and integration of IT, in order to attract students and new faculty, while appreciating the daunting task for existing faculty to keep pace and develop new teaching practices. Considering that the promotion and tenure process is a series of decisions that involves faculty colleagues, academic administration, and the top university administrator, this process may be foiled at any phase.

As Schell explains:

- The chief academic officer may be a proponent of online classes, but someone earlier in the decision process, such as a department chair, may feel that the development of online course materials is of little value. If the department chair finds the faculty member's record insufficient to be passed along for promotion and tenure, then the chief academic officer will not have an opportunity to champion the faculty member's case.

The time demanded in developing new educational technical materials might not necessarily be fruitful when faculty are being considered for promotion and tenure under traditional procedures:

- The lack of a significant relationship between research commitment and computer adoption for instruction weakens the frequent argument that the low value given to teaching in promotion and tenure decisions may be responsible for faculty not wanting to involve themselves in integrating computers into their teaching.

Hirschbuhl *et al.* designed a survey to explore the attitudes and fears of faculty adoption of computer technology and determined that asking the correct questions is essential. The question is not "Why aren't computers widely used for instruction?" The question is "What kind of instructors use computers for instruction?"

The survey intended to ultimately support efforts to build higher education faculty and researchers' confidence and address fears of technology consisted of the following questions:

- To what extent is faculty using computers in their instructional activities?
- Do the personal attributes of discipline, rank, research commitment, and gender relate to faculty's level of adoption of computers for instruction?
- Do organizational factors such as instructional policy, incentives, technical support, and staff development affect faculty adoption of computers in their instructional activities?
- Do attitudinal factors such as computer self-efficacy, computer utility beliefs, and general attitudes towards computers affect the adoption of computers for instruction?

The findings of the study revealed that a consistent correlation associated with faculty that readily incorporated technology in their institutions related to whether the faculty's discipline was quantitative in nature and more adaptive to technology. Otherwise, a more deliberate initiative to incorporate computers and technology was evident.

Another issue surfaced concerning motivations and incentives for developing commuter competence. Institutional and technical support for technology development did not necessarily promote technical competence. However, a valid point was raised that perhaps institutional support did not necessarily motivate development, but instead maintained technical levels of faculty computing.

Evidence from the study indicates that faculty who were aware of technical training that was offered had, in fact, participated in these development opportunities. An unexpected finding of the survey indicated a weak correlation between institutional incentives and the incorporation of computing practices, which may not speak well of the incentive schemes themselves.

As established, the promotion and tenure review process is a series of decisions made by a series of individuals engulfed in institutional missions and philosophies. How institutions and individuals who comprise promotion and tenure committees consider technology and scholarly achievements leaves room for speculations and uncertainty among nontenured faculty in traditional institutions of higher education.

One such individual, Randy Bass, Assistant Professor of English at Georgetown University, who explored and integrated technology with teaching, knew that many tenure—and promotion committees would look askance at the kinds of choices he had made—both because he had spent so much time on new technology and because he had focused on teaching instead of traditional research. Professor Bass favourably convinced the committee that his technology portfolio that provides hyperlinks to the online elements of his introductory course on American literary traditions was a legitimate academic accomplishment and he was awarded tenure as a result.

Professor Bass maintains that his current focus is to encourage other tenure track colleagues, as well as administrators, to now appreciate how teaching with technology is and should be evaluated as a scholarly endeavor.

In this regard, the education of the institution in terms of remaining sensitive to the relationship between the faculty professional advancement process and supporting academic technical initiatives is essential. Gary Bradshaw, tenured Associate Professor of Psychology at Mississippi State University, was warned by colleagues about his ePsych, an online resource for psychology students, which eventually employed eight students and staff and was awarded $400,000 in grants from the National Science Foundation.

Twice he was turned down for a promotion to full professor, where the underlining message resonated that ePsych was not a research project, but was considered more akin to service work.

"As candidates for jobs and promotions stock their portfolios with Internet-related accomplishments, many evaluation committees are skeptical". Across the country it is heard from faculty that the tenure and promotion process often fails to recognize and appreciate teaching with technology, while many institutions are excited and eager for faculty to use and develop online course materials.

"But that might be changing, as a growing number of institutions are working to include digital creations in the tenure folders that form the core of candidates' professional portfolios".

This consists of a paradigm shift in terms of the acceptable format of scholarly work presented for promotion and tenure committee review.

This shift would include the "three reasons selected most frequently for creating technology-based projects are:

1. Instructional impact,
2. Professional recognition, and
3. Motivation for end user".

Identifying the problems relating to specific areas of faculty's technical incompetence and reluctance to become involved with technology-based projects are numerous.

The literature indicates that institutional and personal attitudes, lack of ability and training, areas of discipline, time and workload allocation, lack of appropriate consideration given to teaching with technology during professional evaluation for promotion or tenure are all viable rationale to support technically recalcitrant academic faculty.

The problem for many faculty members becomes quite basic:

- Most of us find it uncomfortable—all right, downright painful—to be in a situation in which we don't really understand what we're expected to be doing. In fact, a power scenario in which our students know more about technology than we do is downright unthinkable.

4

New Communication Technologies

New communications and information technologies have created a global revolution in communications, access to information, and media delivery. These new communications and information technologies are facilitating the sexual exploitation of women and girls locally, nationally and transnationally. The sexual exploitation of women and children is a global human rights crisis that is being escalated by the use of new technologies.

Using new technologies, sexual predators and pimps stalk women and children. New technical innovations facilitate the sexual exploitation of women and children because they enable people to easily buy, sell and exchange millions of images and videos of sexual exploitation of women and children.

These technologies enable sexual predators to harm or exploit women and children efficiently and, anonymously. The affordability and access to global communications technologies allow users to carry out these activities in the privacy of their home.

The increase of types of media, media formats, and applications diversifies the means by which sexual predators can reach their victims. This stage will not attempt to categorise all the types and uses of this new technology. However, this stage will describe the most common and newest of these technologies, and how they are used for the sexual exploitation of women and children.

New and Old Technologies Combined

Older technologies, like television and cable are now combined with modern technologies to create new ways of delivering information, news, and entertainment. Web TV combines the television with the Internet.

New cable networks use satellite transmission to deliver hundreds of channels and pay-per-view delivers content on demand. Presently, there is a high demand for pornographic videos, through mainstream communication networks such as cable TV. Only one of eight major cable companies in the United States does not offer pornographic movies. Satellite and cable companies say that the more sexually explicit the content the greater the demand. ADULT VIDEO NEWS reports that pornography offerings on TV by satellite or cable are increasing video store sales and rentals, not decreasing them, as might be expected. The explanation is that pornography on TV is advertising pornography and finding new buyers.

The mainstreaming of pornography is increasing the exploitation or abuse of women and children used in making pornography.

Owner of ADULT VIDEO NEWS, anything sells:

- There are so many outlets that even if you spend just $15,000 and two days – and put in some plot and good-looking people and decent sex – you can get satellite and cable sales. There are so many companies, and they rarely go out of business. You have to be really stupid or greedy to fail.

Another producer said:

- [A]nyone with a video camera can be a director – there are countless bottom feeders selling nasty loops on used tape. Whatever the quality or origin of a product, it can at the very least be exhibited on one of the 70,000 adult pay Web sites, about a quarter of which are owned by a few privately held companies that slice and dice the same content under different brands.

As a result of the huge market on the web for pornography and the competition among sites, the pornographic images have become rougher, more violent, and degrading. One producer claimed that there were "no coerced" performances in pornography videos, although she immediately acknowledged that "there are little pipsqueaks who get their disgusting little videos out there."

The "misogynistic porn" this producer refers to involves degrading images, such as ejaculation on the woman's face, real pain, and violence against women that results in physical and emotional injuries. In the last ten years, some American and European pornography producers have moved to places such as Budapest, Hungary because of the availability of cheap actors from Eastern and Central Europe. Budapest is a destination and transit city for women trafficked from Ukraine, Moldova, Russia Romania, and Yugoslavia.

There are hundreds of pornographic films and videos produced each year in Budapest. Budapest is now the biggest center for pornography production in Europe, eclipsing rivals such as Amsterdam and Copenhagen. Most Western European producers of sex videos use Eastern European actors whenever possible. An executive at Germany's Silwa production company explained: "They cost less and do more. Even excruciating or humiliating acts usually cost only two or three hundred dollars."

The postal service, traditionally the most anonymous and popular way to transmit pornography, is still used by collectors and producers of child pornography to distribute the pornography. Now, sending materials through the mail is combined with internet technology. Raymond Smith, U.S., Postal Inspection Service, who handles hundreds of cases of child pornography, has found that the rise in internet use by sexual predators has also increased their use of the U.S., mail. He said that from the time they first started investigating child pornography in the early 1980s until five years ago, they have almost eliminated the distribution of child pornography. But since the internet became publicly available, the number of cases connected to the internet has steadily increased. In 1998, 32 percent of cases were related to internet.

In 1999, 47 percent were internet related, and in 2000, 77 percent of the child pornography cases were internet related. Producers of child pornography advertise their videos on the internet and distribute them through the mail. Men in chat rooms trade small files, still images and short movie clips on the internet, but longer movies are sent by mail. Stalkers talk to children in chat rooms, ask them to take pictures of themselves, and send them through the mail. When stalkers convince children to travel to meet them, they send them bus and plane tickets through the mail. Scanners and video digitisers are used to turn old pornographic images, films, and videos into electronic formats that can be uploaded the internet.

About half of the child pornography online is old images from films and magazines produced in the 1960s and the 70s. Digital cameras and recorders enable the creation of images that do not need professional processing, thereby eliminating the risk of detection. These new types of equipment also make it technically easier for people to become producers of pornography. Digital media formats are not static nor independent.

One format can be quickly converted into another. Videos are still the primary production medium for child pornography, and the still images for the Internet are produced from video captured images. From one video, 200–300 still images can be captured, and then uploaded to a newsgroup or to a web site.

Production of child pornography still combines older methods of production, while using new internet technologies for distribution:

- It is safe to say that the number of manufactures has increased over the years with the availability of new medium. Home development of black and white 35MM film, self-developing Polaroid film, video cameras, camcorders, computer scanners, CUseeme technology and now computer cameras have made child pornography easier and easier to produce and reproduce.

One police analyst noted that prior to the internet the majority of collectors of child pornography were not distributors because duplication technology was not readily available. Now, making copies of image files "involves a few clicks of any computer mouse allowing for effortless distribution." Therefore, collectors of child pornography have quickly and easily become distributors.

COMMUNICATION: THE CONCEPT

The word 'communication' is derived from Latin word, 'communis' meaning commonness of experiences. However, communication cannot be defined through a single definition. Different people perceive it in different ways in different contexts. Communication has heen described as "the transfer of conveying of meaning", "transmission of stimuli,' 'one mind affecting another'; 'the mechanism through which human relations exist and develop, or sharing of experience on the basis of commonness".

Communication involves interchange of meaning among individuals. This occurs mainly through verbal and non-verbal symbols, such as language, gestures - a shrug of the shoulders, a nod, facial expressions and actions. Same cultural context makes communication easier because words essentially represent what members of a society decide it would stand for. The knowledge of these symbols, signs and meanings is essential for effective communication. For example, let us take a simple word like 'chair'. To some, it may be a thing to sit on, however, in certain other contexts, it can denote a desired position. Thus we realise that pattern system of communication are influenced by sociocultural- political and economic contexts. Depending on the environment or surroundings in which communication takes place, it can be defined as a process of sharing or exchange of ideas, informataion, knowledge attitudes or feeling among two or more persons to elicit the desired/intended response.

Elements of Communication

Communication is a dynamic process involving active interaction between sender and receiver and variety of inputs. Effective use of communication in a classroom situation between teacher-pupil or pupil-pupil can accelerate the pace of actions.

Following are the elements of communication:

The Source

Source is the communicator. S/he encodes the purpose in the form of a message, to pass it on to receiver and also decides the medium-channel to use for communication.

The Message

Message may be an idea, information or attitude. It can be purposive or non-purposive. Messages drafted for achieving the specific behavioural objective are purposive. Messages with no intention to influence the behaviour are formed as non-purposive. For effective communication, the messages should be short, precise and in simple language. Clarity of message and style of presentation can enhance the effect of communication.

The Channel

The channel for communication is a medium, a carrier of information from the source to receiver and *vice-versa*. This may be verbal, non-verbal, written, printed, visual, etc. TV, Radio, newspaper, etc., are used as means for mass communication.

The Receiver

The receiver is recipient of the message. In the case of mass media, the readers, viewers and listeners are the receivers. In a classroom situation, the teacher is the source, the message is the instruction/lesson and pupils are the receivers.

The Feedback

Receiver's response to communicator's message and *vice-versa* is termed as the Feedback. Feedback is quick in face-to-face communication. This may be verbal or non-verbal. Receiver's feedback to the communicator becomes a stimulus for him/her. This provides an opportunity to gauge the effectiveness of communication and helps in improving the quality of further communication when needed.

Process of Communication

Communication is a necessary condition for growth and transmission of cultures, the continuity of societies and the effective functioning and control of social groups. The process of communication involves interaction between the communicating individuals. In education communication the response (feedback) evoked in the receiver (learner) becomes a stimulus for the communicator (teacher) to which s/he responds. Thus, in face-to-face communication, the source (teacher) and receiver (teacher-student) are at once both response and stimuli. This process continues between the players (source and receiver) in conversation game in which a number of intervening variables such as individual differences, levels of perception, motivation level, etc., are involved.

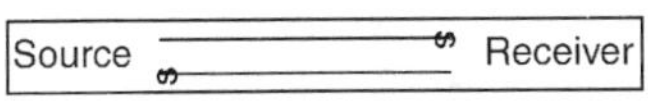

Another important feature of communication is that the recipient (teacher) infers from the behaviour of communicator (pupils), what idea or feeling the other person is trying to convey. S/he then reacts not to the behaviour but to the inferred idea or feeling. The other person then reacts to his/her response in terms of his/her inference of the idea/feeling and the meaning behind it.

Communication accomplishes its purpose accurately if the message is interpreted in the same way by the communicator and by the recipient of the communication. Communication depends on the comprehension and communication skills of the individuals.

THE COMMUNICATION PROCESS

Whenever we ask workshop participants how they would define communication, we hear responses such as "transmitting ideas," "talking and listening," or "sending messages using words and movements." Each person has some notion of what it means to communicate with another and knows how it feels when communication attempts are successful or unsuccessful, yet they have not thought deeply about the communication process itself. People seem to assume that communication works or it does not work, more as a matter of fate than as a process that can be changed or improved. Because communication itself is so complex, we could list multiple elaborate and highly technical definitions of it. For our purposes, however, a simple phrase is an appropriate starting point. As we view it, communication is the symbolic process of sharing meanings. Because this definition is almost deceptively simple, each of these key words needs to be developed.

Symbolic

By saying that communication is symbolic, we mean that symbols are used to transmit messages. Symbols are representations of a person, event, place, or object. Words, or verbal expressions, are the most commonly understood symbols, but symbolic actions also include the whole range of non-verbal behaviours: facial expressions, vocal tone, eye contact, gestures, movement, body posture, appearance, context, and spatial distance. In addition, objects and ideas can be used as symbols. For example, friends often exchange gifts, food, or e-mails as symbols of connectedness. You have learned to use verbal and non-verbal symbols both as a message creator and as a message interpreter. As a speaker you create messages by selecting the most appropriate symbols from a range of options in order to reach your intended receivers most effectively. As a listener, you attempt to interpret the symbols others convey to you. Although exchanging appropriate symbols appears rather simple and straightforward, we are constantly amazed at the communication breakdowns that occur as symbols are misinterpreted. Effective communicators are those who are able to select the most appropriate symbols or messages for specific other persons and who are able to interpret the intended message symbols of other speakers. As a child you learned to encode one type of message to ask your father for money and another to request a loan from your best friend. You learned to interpret your brother's gestures indicating if he is feeling sad, worried, or exhausted. You have learned who will be enraged if you roll your eyes at them, and who will decode your non-verbal cue as humorous. For effective communication to occur, the speaker and listener must share the same meanings for the symbolic messages they exchange.

Process

Relational communication involves a process, a dynamic and continuous process. Each relationship develops its own communication history, a history that cannot be rewritten. Someone once said, "It's unfortunate you only get one chance to make a first impression." A relationship's communication pattern begins at the first moment of contact— at a party, in a classroom, in a meeting. The relationship may start in any number of ways: a question, a glance, an introduction, or a smile. Once contact is made, the relationship begins to develop its history, which is constructed and reflected by its communication patterns.

A relationship's ongoing development may be interrupted by physical or psychological distance; relational partners may move in and out of each others' lives over many years, but the history of the relationship continues from that first meeting. Sometimes people say they wish to wipe out a time period of their relationship or forget a painful argument that occurred. Individuals may choose to emphasise or deemphasise certain communication events throughout their relationship history, but they can never go back to "how things used to be," or delete a piece of that history. Although some models of the communication process portray it as a circular process, our preferred model for understanding this process comes from the thinking of Frank Dance, whoproposed a helical representation. Imagine the form of a helix, in which the continuousness of the process is represented by the infinity sign. This model depicts the ever-widening scope of the relationships as participants continuously reencounter each other, a process that continues indefinitely.

Whereas a circular model suggests that communication returns to the same place, the helical model implies the ever-changing, progressive, and evolving nature of relational interactions. The helix representation provides support for the concept that "you can't put a relationship into reverse and erase a difficult period of time."We have stopped counting the times workshop participants say, "If only things could go back to the way they were two years ago," or, "I want to wipe out the last six months of our marriage." In reality each encounter has inalterably added to their relationship, and this history cannot be denied.

People in relationships cannot wipe out a huge hurtful fight, long periods of verbal aggression or silence, or, in romantic situations, the affair.Yet most friends, partners or colleagues can learn to manage their history in effective ways: through emphasising positives, talking through the conflicts, and behaving in ways that affirm their ties. A conflictual father and son cannot pretend they never hurt one another with words or fists; friends cannot erase sarcastic comments. All they can do is work through the issues that currently keep them from dealing with each other in constructive or caring ways, and attempt to change their present communication patterns as they continue in their relational process. People can always choose to change, to do things differently. Such choices represent one of the most exciting parts of the relational development process.

Sharing

Even though the words speaker and listener are commonly used in communication terminology, communication is not a process of trading messages. It does not resemble a poor tennis match in which one Player A hits the ball and then just stands there until the Player B hits the return. Symbolic messages do not travel from Person One to Person Two and back to Person One again in some turn-taking ritual. Rather, at its most basic level, communication requires mutual and continuous involvement, sometimes referred to as the transactional nature of communication. This mutual influence process is similar to a skilled tennis match in which both players are always in motion based on what they anticipate the other will do. Similarly, in communication encounters both parties remain actively involved in the process. For example, even though Michael may appear more talkative, Vanessa conveys non-verbally that she is bored or pleased or annoyed, thus influencing Michael's choice of future message symbols. Both are actively and continuously involved in every moment of the interactions; thus, the mutual influence process that characterises interpersonal interaction.

As relationships develop over time, the transactional process becomes more complex. Your perception of another person and that individual's perception of you combine to form a context for your interactions. If you see Person X as warm and supportive, you will relate to him or her in an open manner. Person X is then likely to see you as open and friendly and relate to you with increasing warmth or support. Thus, your perceptions of each other affect each interaction as well as the overall perception of the relationship. The situation can also be reversed, creating a negative context. If you see another person as judgemental or sarcastic, this may lead you to interact in a defensive or combative manner.

You may be caught up in a type of negative spiral. Each communication exchange occurs within the context of a mutually constructed relationship. If the definition of a relationship remains relatively unchanged, for example boss and employee, romantic partners, and parent and child, the nature of the communication process becomes fixed. Each new encounter reinforces the good or the bad. A boss who constantly relates to staff members as incompetent may stifle their attempts to be innovative. A parent who treats children as responsible persons fosters their ability to handle new situations. This process is captured in the statement "Over time we create an image of another person and relate to the image we create." Individuals construct a reality of themselves and of others through their interactions, and relate to those realities they constructed. The attempt to understand and adapt to another represents a communication challenge.

Meanings

Although verbal and non-verbal symbols permit us to transmit thoughts and feelings, the symbols must be mutually understood for the meanings to be truly shared. Common meanings make it possible for us to communicate. Since there is no absolute standard for all symbols, we are constantly trying to connect with people, even our family members, who do not share exactly the same meanings for the symbols that we use. Therefore it is important to remember the expression "Words don't mean; people do."

Each person's background, including physiological state, family and cultural background, and unique experiences, influences how he or she perceives the world and attaches meanings to symbols. The experience of being nearsighted, athletic, extraverted, dyslexic, artistic, or shy affects how you perceive the world and relate to others. Your family of origin served as your first communication classroom, teaching you how to interpret messages and how to use communication to manage key relational issues such as intimacy and conflict. In addition, your cultural background, socioeconomic level, and educational experiences influence your perceptions. Based on your culture, you may interpret big hugs, multiple-course meals, and shouting voices as symbols of caring. If you grew up in a lower-middle-class neighbourhood, you will have different meanings for money and security than someone who grew up in an affluent community. Finally, your own unique circumstances influence how you assign meanings.

A painful custody battle affects how you discuss divorce. Early school experiences influence how you participate in college classes. Living abroad affects how willing you are to interact with people of different cultures.

Fortunately, most people report many similar experiences, but no two people develop the same set of meanings. Each is a unique entity with particular meanings for certain symbols. Ninety dollars may represent a large purchase to one person, but her partner assumes that only purchases above $500 are large. A nickname may seem funny to you and insulting to your friend. Screaming may be viewed as an acceptable or terrible way to resolve conflicts. Breakdowns in communication often occur because of missed meanings. Only with knowledge and empathy can you walk in someone else's shoes, experience the world from a different perspective, and create messages that reflect that point of view.

Frequently, focus is placed on the words rather than on the entire range of verbal and non-verbal symbols that are constantly being used to create and interpret the meanings of messages. Therefore, at any point in time each person involved in communication is contributing to the process—and experiencing the transactional nature of the communication process.

Effective communication requires the psychological presence of both parties—attention and connectedness are indicators that both parties are focused on the encounter. We find the following simple exercise very useful for demonstrating how individuals may differ in translating the meaning of everyday terms. Imagine yourself saying these phrases to a particular person. Think about exactly what you would hope that other person would do if you said these words to him or her.

Each of the italicised words receives many different responses. Depending on the person responding, more respect may be indicated by listening to me, asking for my opinion, or following my advice. Trust may be indicated by keeping what I say confidential, telling me your real feelings, or telling me when a painful event happens. These are only examples of the many meanings that people have for the two common terms respect and trust. Shared meanings are critical because they help to create the context for a relationship in which participants learn to predict how the other will react to particular verbal and non-verbal messages.

THE DEVELOPMENT OF INFORMATION TECHNOLOGY

The key technology in micro-electronics was the transistor, invented in 1947 at Bell Laboratories in Murray Hill, New Jersey, by three physicists, Bardeen, Brattain, and Shockley, who were later awarded the Nobel Prize. Other key technologies were the use of silicon as a medium (1954) and the integrated circuit (1957). Together these three made the development of semi conductors and microchips possible, and the price of these dropped very rapidly in the years that followed (85% between 1959 and 1962, and from $50 dollars for an integrated circuit in 1962 to $1 in 1971). The microprocessor chip was invented by Ted Hoff of Intel in 1971, and these gradually became more and more powerful.

Computers really began with the first general purpose computer, the ENIAC (Electronic Numerical Integrator and Calculator) in 1946. This was a huge monster, weighing 30 tons and consuming enormous quantities of electricity. Many of the early computers were made by business machine companies, most of which went out of business as the new specialist computer companies appeared, and much of the funding came from the government and the defense industry. Gradually the market settled down into large mainframe computers, and smaller mini computers, and for the most part they worked with punched cards and paper tapes rather than keyboards. Together with screens, they came later.

With the invention of the chip, costs went down together with size, and the power of the machines increased rapidly. By the early 1970s, powerful programmable calculators had begun to appear, and they rapidly became smaller and cheaper. In the mid 1970s, personal computers also started to appear, notably the Apple I, and Apple II, "the first commercially successful microcomputer, realized in the garage of their parents' home by two young school drop-outs, Steve Wozniak and Steve Jobs, in Menlo Park, Silicon Valley, in a truly extraordinary saga that has by now become the founding legend of the Information Age." (A film about this period came out a year or so ago, called 'Pirates of Silicon Valley', mainly about Bill Gates and Steve Jobs). The Apple Macintosh followed in 1984. The company, which started up with capital of $91,000 dollars, had sales of over $500 million by 1982. IBM responded to Apple by producing its own Personal Computers (PC) which were the ancestors of the Windows machines at the present day. It is important to remember that the cost of storing information has halved about every six months since then, and that the desk-top computers soon became as powerful as the mainframes of only a few years before.

Silicon Valley itself is in Santa Clara County, about 50km south of San Francisco, and very near Stanford University. The growth of this rural area as the center of the high-tech industry began with Frederick Terman, the dean of engineering at Stanford. He helped two of his graduate students called Hewlett and Packard start an electronics company in the area in 1938, and the firm became wealthy during the War. In 1954 Stanford hired Shockley, the inventor of the transistor, and he recruited a group of eight bright young engineers, most of them from the Bell Telephone labs. Relations between Shockley and his engineers were not particularly good, and his company collapsed, but the engineers went off and started their own companies, and from these grew half of the 85 largest American semiconductor firms, including

Intel. Stanford replaced MIT as the major center in this kind of research.

The new firms were much more flexible and innovative than the older ones in the East, and they shared information at informal meetings. One of such gatherings was the Home Brew Computer Club, who included Bill Gates, Steve Jobs, and Steve Wozniak, and an article read at a club meeting stimulated Wozniak to design his Apple I machine with Steve Jobs in a garage in 1976. Gates founded Microsoft at around the same time. Now in addition to the original companies, Japanese, Taiwanese, Korean, Indian, and European companies have also moved in. A lot of business comes from Hollywood, with its demand for computer graphics and special effects, so that the computers are just one part of a multi-media industry.

IT and Globalization

The effects of information technology have been to speed up the processes of globalization in the world economy. After the Second World War, the costs of research and development very quickly led to the development of multinational corporations, and with the decreasing cost of transmitting information these have been able to organize their research and production on a global rather than a national or even a regional basis. Three interdependent core areas of the world economy have emerged – North America, Europe and East Asia. The rise of information technology is also linked to the development of a global currency market, based in the three world cities of London, New York and Tokyo, and the rise of financial services industries which are also based there as well. (This is the theme of Sassen's book mentioned earlier.)

Deregulation of the currency markets mean that capital can be transferred anywhere in the world electronically, and there is a vast subsidiary industry of people making money through betting on future currency movements. The result has been to amplify instability in world markets, and of course to marginalize even further the areas and people outside the reach of this new technology. And it is the corporations which have flexibility and which are able to adapt quickly which have been particularly successful in this kind of environment.

The East Asian corporations were particularly well placed to take advantage of this new technology, because of the nature of the links between them. The Japanese industries were organized around the *keiretsu*, networks of companies both big and small, with one or other of the large banks at the center. Along with the growth of the giant corporations has gone the 'small is beautiful' philosophy, and Japan combines the two ideas, with huge companies sitting at the top of a hierarchy of small family businesses. The Chinese businesses in East Asia are linked together in a similar way through family ties, which give flexibility in terms of flows of capital, loans, information, and so on. But the network enterprise is also marked flexibility of employment, the collapse of traditional work patterns in many cases, and the massive influx of women into the labor market – often at lower rates of pay than their male counterparts. In the UK this took place during the Thatcher period of the 1980s, and in Japan it still seems to be taking place at the moment.

In the second volume of *The Information Age, The Power of Identity,* Castells looks at some of the social impacts of these changes. The shift of power in the work place, the breakdown of traditional patterns of employment and the gender division of labor, has, Castells argues, produced a crisis in the family. 'Normal' families with husband, wife and their children living together have now become the minority rather than the norm, thanks to the increasing divorce rate, and alternative lifestyles.

The reaction has been in many cases religious fundamentalism – one of the attractions both of fundamentalist Islam and Christianity is that they seem to restore the balance, with a heavy emphasis on the integrity of the family and patriarchal dominance – the woman's place is in the home. Many of these groups are also opposed to what they see as the New World Order, dominated by the United Nations and similar international organizations, so that much of the resistance to globalization is coming from ethnic, extreme nationalist and religious groups. Resistance to the new global order has resulted in the flowering of a diversity of cults and social movements, from fringe cults such as AUM Shinrikyo in Japan to environmental movements of various types both regionally and world wide, and there are lengthy discussions of both of these.

Finally, of course, the state itself has been weakened by these processes of globalization. Increasingly they are unable to stop flows of capital, or flows of information, or to prevent the major corporations from moving industrial production to other regions of the world. Democratic processes have been profoundly influenced by television, especially in the United States, so that real debate has been replaced by the production of 'sound bites' and video clips endlessly repeated on television talk shows, and the politics of scandal, reducing the political process to the format of a soap opera.

The final volume, *End of Millennium,* returns to the global economy, and looks at the role of the state in the economic process.

The Collapse of the Soviet Union

There is a lengthy discussion of the collapse of the Soviet Union, which Castells argues had actually been more successful economically than is often thought nowadays. The Soviet Union actually survived a series of major disasters, from the civil war after the Revolution and the purges of the Stalin period to the Second World War (in which 25 million Soviet citizens died) –

and still managed to re-house almost the entire population and gain a technological lead over the US in space research by the mid-1950s.

The collapse of the Soviet Union, Castells argues, came about not because socialism itself was inherently flawed, but because the Soviet version of it, with heavy state control, was unable to manage the transition to the new informational economy. This was because the economic growth which made victory in the Second World War possible was based on state planning which was largely insulated from the rest of the world and which could not adapt to the new conditions. Much of the production was in any case for defence, and did little to increase the standard of living of the average citizen.

Despite its successes in space, and despite producing some of the best mathematicians in the world, the Soviets fell further and further behind in information technology. They could design theoretical chips, but lacked the facilities to actually make them! The crucial decision was to base their computing systems on IBM technology which by the 1960s had become the best in the world, rather than develop their own systems. (One of the reasons for this was that the Soviet generals wanted the best available hardware for the military.) From then on the Russians were dependent on increasingly obsolete systems borrowed or pirated from the West, and they were soon 20 years behind in computer design. Even copying and typewriter facilities were tightly controlled for political reasons.

Finally, the growth of the system meant that there was a privileged class of officials who benefited from it – and who resisted change. In the 1980s, Gorbachev tried to reform the system with the two famous catch words of *perestroyka* (retructuring and *glasnost* (openness) – and an attempted counter-coup by senior officials in 1991 led to the collapse of the Soviet Union itself.

Global Crime

In the aftermath of the collapse, organized crime has flourished, organized partly by former party officials, especially in the republics of the Caucasus and Central Asia. Castells and his students have also carried out a good deal of research into the drugs trade in Latin America, and the various cartels which organize it. One of the features of organized crime world wide is the increasing integration of the various regional mafias which control it, and the amounts of money in circulation are staggering. The whole system incidentally is based on the criminalization of drug taking in the US – it seems that the government learned nothing from the prohibition experience during the 1920s which established the Mafia with a power base in terms of wealth and influence which they have never lost. Ban something which is in heavy demand, whether it be booze, drugs or sex, and the result is a black market and organized crime.

The 'Fourth World'

Much of the attraction of criminal activity lies in the fact that there are few alternative sources of income for many people, especially in what Castells calls the 'fourth world'. These are the regions and peoples excluded for one reason or another from the new information economy. It includes whole regions – such as Subsaharan Africa – where the increasing struggle for survival has led to the collapse of economies and regimes and an upsurge in inter-ethnic violence, resulting in almost perpetual civil war in states such as Liberia, Sierra Leone, Zaire, Somalia, Rwanda and Burundi.

But it also includes the excluded minorities in the core industrial countries, especially the United States. One of the features of the labor market resulting from informationalization is polarization – while the rich are getting much richer, the majority, even in the United States, are actually getting poorer, and an increasing proportion have literally no assets at all. Things are worst in the ghettos, with high rates of crime – so that a large proportion of the younger men are either in prison or on probation. More hopeful is his discussion of economic growth in East Asia, but even here clearly the statist model which brought about economic growth in the postwar period, first in Japan and then in the four little tigers and China itself, is under strain from the informational economy as we move into the 21st century.

ADVANCES IN INFORMATION TECHNOLOGY

The IT revolution drives the extraordinarily rapid decline in the cost and rapid increase in the processing power of digital technologies. The digital device whose technological advance has perhaps been most crucial to the IT revolution is the microprocessor, the collections of millions of tiny circuits that serve as the "brains" of personal computers and that are embedded in an ever-expanding number of products, from video games, to cars, to refrigerators. Over the past two decades, the processing power of microprocessors has doubled roughly every six months.

Rapid advancements in fiber optic technologies have also been critical to the IT revolution. Fiber optics technology enables data, including voices captured in digital form, to be converted into tiny pulses of light and then transmitted at high speeds through glass fibers wrapped into large capacity telecommunication cables. Hundreds of thousands of miles of these cables were installed over the past ten years, boosting the speed and capacity of telecommunications networks. A contributing factor to the growing technology sector is human capital.. The majority of tech firms worldwide have levelled the baseline production of new technology to the point where they seek new areas of improvement for their products. Human capital, the workforce, drives these advancements and often the reason why one company succeeds, while others do not. Tech firms seek skilled workers with knowledge of technology and problem solving skills, which gives them an edge over the competition. Technology companies in the U.S. are pushing for better immigration policies so they can hire the best and the brightest from around the world.

The transformation of the technology sector in the U.S. market resulted in need for software developers, computer and information systems managers, and computer systems analysts. New jobs such as these are commodities in the globalized world of technology, especially for companies recruiting individuals from technologically advanced countries. The growing market for tech jobs will continue to increase as technologies become even further integrated into society.. More and more jobs will become available to individuals that obtained degrees in technology orientated fields. According to Catherine Mann (2003) "Frequently cited projections indicate that millions of jobs will be lost to offshore workers. What these projections ignore is that the globalization of software and IT services, in conjunction with diffusion of IT to new sectors and businesses will yield even stronger job demand in the United States for IT-proficient workers."

Driving Down the Cost of Information Transactions

A key reason why these advances in IT have spread so quickly is that they have progressively reduced the unit cost of computing power or the transmission of a message. For less than $400, Americans without any advanced technical training can purchase and use a desktop computer whose data processing power far exceeds the room-sized computers that powered the spacecraft that carried astronauts to the moon and back in the late 1960s and early 1970s. Companies such as Microsoft have even sold $100 computers to consumers in emerging countries as a way of helping developing countries use more advanced technological resources.

The decline in computing prices is a factor in spurning the growth of computers in the developing world. Countries such as Zimbabwe, India, Brazil, and China experienced tremendous growth in the number of personal computers. From 2000 to 2005, the growth rate of personal computers per capita exceeded 200 percent for most of these nations, with developing countries like Zimbabwe increasing almost 600 percent. By the end of 2011, there were reportedly more than 1.6 billion PCs in use, with the U.S. in first place and China in second.

The spread of digital technologies has also been spurred by several unique attributes of information, which serves as the principal input and product of many IT industries. In contrast to more tangible products, like consumer goods, one person's "consumption" of a piece of information does not necessarily reduce or eliminate the possibility that another person might benefit from the same piece of information.

Furthermore, networks built upon the exchange of information, like the Internet, tend to become more valuable to existing participants as new participants link up with them. Finally, the cost of using digital technologies, such as Internet service providers, decreases as the number of users increases. All of these factors have worked together to promote rapid growth in the demand for, and supply of, IT products and services.

During the second half of the 1990s, as more people bought computers and went online, the average cost of the equipment and services necessary to access the Internet declined. Today, individuals go beyond the conventional desktop computer to stay connected: Wi-Fi networks, laptops, smartphones, tablet PCs, and even phones utilize Wi-Fi networks to make the Internet an integral—and necessary—part of everyday life.

INFORMATION TECHNOLOGY AND YOUR LIFE

This stage is about computers, of course. But not just about computers. It is also about the way computers communicate with one another. When computer and communications technologies are combined, the result is *information technology*, or 'infotech.' *Information technology (IT)* is a general term that describes any technology that helps to produce, manipulate, store, communicate, and/or disseminate information. IT merges computing with high-speed communications links carrying data, sound, and video. Examples of information technology include personal computers but also new forms of telephones, televisions, appliances, and various handheld devices.

Distinguish Computer Technology and Communications Technology

Computer Technology

You have certainly seen and, we would guess, used a computer. Nevertheless, let's define what it is. A *computer* is a programmable, multiuse machine that accepts data—raw facts and figures—and processes, or manipulates, it into information we can use, such as summaries, totals, or reports. Its purpose is to speed up problem solving and increase productivity.

Communications Technology

Unquestionably you've been using communications technology for years. *Communications technology*, also called *telecommunications technology*, consists of electromagnetic devices and systems for communicating over long distances. The principal examples are telephone, radio, broadcast television, and cable

TV. In more recent times, there has been the addition of communication among computers—which is what happens when people 'go online' on the internet. In this context, *online* means using a computer or some other information device, connected through a network, to access information and services from another computer or information device. A *network* is a communications system connecting two or more computers; the internet is the largest such network. Information technology is already affecting your life in exciting ways and will do so even more in the future. Let's consider how.

Information Technology through Computers

These days, it seems as if everything is controlled by computers. Not in a bad way, of course: Stanley Kubrick's worst fears as witnessed by *2001: A Space Odyssey* won't be coming true anytime soon. But the fact remains that from the databases maintained by the IRS to the purchasing records of supermarkets to the information doctors keep of their patients, computers have become the single essential tie that binds the world together. It therefore should come as no surprise that a degree in information technology is more marketable now than it ever has been in the past.

Information technology "is the study, design, development, implementation and support of computer-based information systems to address real -world problems". In other words, it is the science (and, indeed, some might even say the art) of working with computers in order to facilitate the easier and more efficient use of them by non-professionals and professionals alike.

Niche Areas: As has been noted several times already, there are many specific areas of the IT field on which you may choose to focus. They include, but are not limited to, the following:

- Graphics and Imaging
- Operating Systems
- Personal Information Management
- Publishing
- Document Management
- Statistical Analysis & Mathematical Modelling
- Document Integration
- Administrative Systems

Types of Associate, Undergraduate & Graduate Programmes Available: Over the course of the past decade, more and more colleges and universities have begun offering degree programmes in information technology. This is because of the ever-increasing importance of the skill-sets IT practitioners possess. An associate's degree in IT will prepare you for further study at the undergraduate level, as well as provide an adequate base of knowledge that you may attempt to parlay to a job as an entry-level programmer or database manager.

A bachelor's degree in IT affords you the opportunity to learn about all aspects of the field and to begin the process of narrowing down your areas of specific interest so that you can attain a certain amount of expertise. Graduate degrees in IT are generally either for those who wish to pursue collegiate teaching or research, or who wish to eventually work in the highest levels of the technology world—government computer development or computer and software design, for example.

Typical Admissions Requirements: Getting into an associate's degree programme in IT is not terribly difficult at all. As long as you have an interest in the field and a bit of previous experience with computers, then you should be fine. At the higher levels of education in the IT field, however, competition can become rather stiff.

Indeed, because of the high skill level of many applicants and the popularity of the field in general, there are likely to be many people with high levels of ability vying for a small number of spots. Don't let that discourage you, however. If you are truly interested in the field and even if you don't get in the first time you try, you should not throw in the towel. Perhaps with a bit of work experience and a bit of additional effort, you, too, will be on your way to a degree in information technology.

Careers in This Field: Because of the many and wide-ranging applications of computers and software, there is a nearly infinite variety of career options available to graduates of IT programmes. They include, but are not limited to, the following:

- Computer Software Engineer, Systems Software
- Computer Systems Analysts
- Computer Support Specialists
- Computer Programmers
- Computer and Information Systems Managers
- Computer Software Engineers, Applications
- Database Administrators
- Computer Security Specialists
- Computer and Information Scientists, Research

Salary Ranges in this Field: Because of the wide variety of careers available to those possessing degrees in IT, it is difficult to narrow down how much they stand to make. A government employee, for example, may very well make less money than an IT expert working with, say, a publicly-held financial institution on Wall Street. Also affecting the amount of money you stand to eventually make is the specific area of IT on which you choose to focus. Therefore, in order to more fully and accurately gauge what you'll eventually earn, it is best to do research on as specific an area as possible in the IT field.

Future Outlook: Because the IT field is so large and wide-ranging, it is virtually impossible to discuss the job outlook in general terms. However, the field of computer and information systems managers is as good a place to begin as any and the outlook for it is likely indicative of larger trends in the field. According to the United States Bureau of Labour Statistics:

"Employment of computer and information systems managers is expected to grow faster than the average for all occupations through the year 2014. Technological advancements will boost the employment of computer-related workers; as a result, the demand for managers to direct these workers also will increase. In addition, job openings will result from the need to replace managers who retire or move into other occupations. Opportunities for obtaining a management position will be best for those with computer-related work experience; an MBA with technology as a core component, or a management information systems degree; and strong communication and administrative skills.

Despite the downturn in the technology sector in the early part of the decade, the outlook for computer and information systems managers remains strong. To remain competitive, firms will continue to install sophisticated computer networks and set up more complex Internet and intranet sites. Keeping a computer network running smoothly is essential to almost every organization. Firms will be more willing to hire managers who can accomplish that.

Similarly, the security of computer networks will continue to increase in importance as more business is conducted over the Internet. The security of the Nation's entire electronic infrastructure has come under renewed scrutiny in light of recent threats. Organizations need to understand how their systems are vulnerable and how to protect their infrastructure and Internet sites from hackers, viruses and other acts of cyber-terrorism. The emergence of cybersecurity as a key issue facing most organizations should lead to strong growth for computer managers. Firms will increasingly hire cybersecurity experts to fill key leadership roles in their information technology departments because the integrity of their computing environments is of utmost concern. As a result, there will be a high demand for managers proficient in computer security issues.

With the explosive growth of electronic commerce and the capacity of the Internet to create new relationships with customers, the role of computer and information systems managers will continue to evolve. Persons in these jobs will become increasingly vital to their companies. The expansion of the wireless Internet will spur the need for computer and information systems managers with both business savvy and technical proficiency."

History of Information Technology

The term "information technology" came about in the 1970s. Its basic concept, however, can be traced back even further. Throughout the 20th century, an alliance between the military and various industries has existed in the development of electronics, computers and information theory. The military has historically driven such research by providing motivation and funding for innovation in the field of mechanization and computing.

The first commercial computer was the UNIVAC I. It was designed by J. Presper Eckert and John Mauchly for the U.S. Census Bureau. The late 70s saw the rise of microcomputers, followed closely by IBM's personal computer in 1981. Since then, four generations of computers have evolved. Each generation represented a step that was characterized by hardware of decreased size and increased capabilities. The first generation used vacuum tubes, the second transistors and the third integrated circuits. The fourth (and current) generation uses more complex systems such as Very-large-scale integration or System-on-a-chip.

Information Technology Today

Today, the term Information Technology has ballooned to encompass many aspects of computing and technology and the term is more recognizable than ever before. The Information Technology umbrella can be quite large, covering many fields. IT professionals perform a variety of duties that range from installing applications to designing complex computer networks and information databases. A few of the duties that IT professionals perform may include:

- Data Management
- Computer Networking
- Database Systems Design
- Software design
- Management Information Systems
- Systems management

Information Technology Title

The Commonwealth of Pennsylvania has made a substantial investment in information technology resources. Many state agencies have installed a mainframe computer, midrange computer and a local area network. Our computer systems and programming languages are continuously being evaluated and updated. Also, information technology specialists receive training in the latest technological advances.

State agencies have developed several dynamic and innovative training programmes to introduce new employees to their systems. Classroom training, specialized workshops and vendor courses form the core of the training. Employees have the opportunity to follow a progressive career path that can lead to higher levels of computer programming or computer systems analysis and management.

Many of the latest concepts in the computer field are under development or nearing implementation including integrated communications networks, client server technology, distributed technology, imaging technology, CASE tools, computer assisted design, artificial intelligence and geographic information systems. Pennsylvania state government has also been recognized as a national leader in its' management of the Year 2000 date modification. In addition, the Commonwealth is committed to providing information to the residents of Pennsylvania by utilizing the technological opportunities available through the Internet.

Information technology (IT) is a label that has two meanings. In common usage, the term "information technology" is often used to refer to all of computing. As a name of an undergraduate degree programme, it refers to the preparation of students to meet the computer technology needs of business, government, healthcare, schools and other kinds of organizations.

IT professionals possess the right combination of knowledge and practical, hands-on expertise to take care of both an organization's information technology infrastructure and the people who use it. They assume responsibility for selecting hardware and software products appropriate for an organization. They integrate those products with organizational needs and infrastructure and install, customize and maintain those applications, so providing a secure and effective environment that supports the activities of the organization's computer users.

In IT, programming often involves writing short programmes that typically connect existing components (scripting). The shaded area extends down most of the right edge as it focuses on the application, deployment and configuration needs of organizations and people over a wide spectrum. Across this range (from organizational information systems, to application technologies and down to systems infrastructure), the role of the IT specialists has some overlap with IS, but IT people have a special focus on satisfying human needs that arise from computing technology. In addition, the IT shaded area goes leftwards from application towards theory and innovation, especially in the area of application technologies. This is because IT people often develop the web-enabled digital technologies that organizations use for a broad mix of informational purposes. This implies an appropriate conceptual foundation in relevant principles and theory.

Technology, Telecoms and Contact Centres

Over the last few years, these industries have been among the fastest growing and developing sectors in the country. They have an impact on all industries.

- About two per cent of all Scottish jobs are in the IT and telecoms industries. (48,700)
- There are also IT professionals employed in other sectors; over half a million in the UK.
- The UK is the third largest telecoms market in the world.
- There are 56,000 contact centre jobs in Scotland; almost 86 per cent of these are in the central belt of Scotland.
- Most industries have contact centres; financial services, with 30 per cent of the total number of jobs, remains the most important user of contact centres.

Areas Included: The IT and telecommunications sector is responsible for:

- the installation and maintenance of communication networks
- the installation and support for new computer software and hardware

These industries have changed the way we lead our lives and altered the way in which business is conducted; you can play videos on mobile phones and hold face to face meetings with business colleagues in other countries.

There is almost no aspect of our lives that is not affected by this industry.

In Contact Centres, staff no longer just use a telephone; they also use the Internet, e-mail and SMS messaging.

- More than half of all jobs are both professionals and associate professionals.
- The most common IT professional job role in the UK is "software engineer".
- Contact Centre staff require high levels of technical, language and personal skills.

What's it like to work in the information technology, telecommunications & contact centres industry

- This industry will appeal most to those with an interest in technical work.
- Most jobs are full time – 86%.
- Most employees are men – 62%.
- Self employment in IT is slightly higher than the average.
- The average age of a worker is 36 – younger than the average of 40 for Scotland as a whole.

- In contact centres, three fifths of the workforce is female (61%).
- The prospects for graduates are good. In the last two to three years, one in four employers had recruited a graduate.
- Employers are less likely to recruit school leavers. Less than one in ten employers have recruited a school leaver in the last two to three years.

Future Trends

- Employment is expected to increase by more than 20% in the next ten years.
- Future skills shortages are expected to be in systems integration, networking and business analysis.
- Despite some call centres moving overseas and changes in technology; the industry is predicted to continue to grow.

CHALLENGES IN INTEGRATING INFORMATION & COMMUNICATION TECHNOLOGY IN SCHOOL EDUCATION

Technology is rapidly changing in all spheres of life, so in education. It has become very necessary to know how we learn, how we express ourselves, and how we perceive and interact with our world. As a means of responding to the change and surviving in an increasingly competitive educational environment, management institutes must develop a strategic plan for enhancing management education and value adding to the existing system of education. As a means of achieving the aforesaid goal the use of computers and other technology in support of student learning, accommodating the full range of learning styles and special needs, and fully integrating technology into the curriculum has become extremely needful in this hour. A lot of related issues are also involved in such a process as this includes administrative, planning and budgeting; professional development for teachers and administrators; and, very importantly, ensuring equity of access and use. The current emphasis is ensuring that technology is used effectively to create new opportunities for learning and to promote student achievement.

In almost all management institutes the main focus is laid on the employers need who prefer: This has been derived somehow from the student's psyche as success for them is a respectable degree, good education, good job and higher salary, to meet such a demand each management institute should plan strategies to enhance the learning experience of the students so that they get better value of management education and have a better prospect in job market.

Some basic principals of good teaching are never going to change even teaching involves the use of ICT, as this has been found my many researchers that it remains a matter of concern for teachers who are not able to shift from traditional class rooms to a modern technology driven classroom. It is central to the role of a teacher to help pupils learn.

As the nature of what is to be learnt changes, so does the role of the teacher in that process. The teacher was once the sole source of information and was expected to know all the relevant knowledge and to transmit it effectively to the pupils. But now with the advent of various technologies, when we sometimes find that some students do have access to more information than any teacher can ever know, thus it has become very necessary for a teacher to change the role of being an instructor who was considered as a source of all knowledge to a facilitator who by the experiences, knowledge and skills help students, learn, understand, evaluate and communicate knowledge.

The role of the teacher is not being diminished but changed. What is necessary above all is that teachers develop the confidence that they are important to their pupils, not because they happen to know some facts that pupils do not know, but that they have a vision of what they want pupils to be educated in as well as the requisite knowledge, skills and understanding to facilitate and monitor pupils' progress. The fact that pupils may know more about the operation of a piece of software than the teacher can be of benefit to both.

When computers were first brought in as an educational tool, many articles, press releases gave a very fearful idea to the teachers about "teacher less classrooms". But truth is technologies do not do the way it was depicted. There is no simple way to implement new procedures. But if the process is thought through and well handled, the dividends can be huge. As business schools experience increased competitive pressures, information technology is one area that schools might use to differentiate or compete with or, more importantly, to use as a catalyst for transforming educational processes. ICT has definitely proved out to be an efficacious means of enabling intentional changes in teaching and learning processes.

Before discussing the technologies available for it, becomes very important to discuss the learning models and Role of ICT.

Objectivist Model (Yarusso 1992)

This model assumes that the goal of teaching is to efficiently transmit knowledge from the expert to the learner. The lecture method of teaching embeds the pedagogical assumptions of this model. The transfer of knowledge can be by the means of electronic presentation. The softwares which are easily available, like power point etc can be used. If the faculty wants to make the presentation more interesting other softwares like macromedia director (for giving interactive effect) media player (to display some clips related to the topic) etc. can be used.

Constructivist model (O'Loughlin 1992)

It calls for learner-centred instruction: individuals are assumed to learn better when they are forced to discover things themselves rather than when they are told, or instructed. It is found that the learner develops the skill hypothesizing, predicting, posing questions, researching answers, imagining, investigating, and inventing. Following this model ICT's role in imparting knowledge is by providing a particular problem area by the faculty. Rest finding the answer by the students can be by the use of online databases, internet search engines or CD-ROMs or electronic library.

Cooperative Model of Learning (Slavin, 1990)

The major goal of cooperative learning is the construction of shared understanding through interaction with other individuals, an implicit goal is improving communication and listening skills and eliciting participation. One implication of the cooperative model for instructional methods is that the instructor's role is to facilitate maximal information and knowledge sharing among learners rather than controlling the content and delivery of learning. This can be brought about by networked computers where each and every individual learner and instructor is able to view the work of everybody in the classroom.

Higher education is beginning to change in response to providing educational opportunities to an increasing diverse population of students. A new type of university emerged in the last 25 years which John Daniel (1996) thinks will hold lessons for the renewal of all universities. Called "mega-universities".

Prominent Business gurus have given traditional universities just 30 more years before dying off (Cunningham et al., 2002). As radical as their predictions may be, it does require colleges and universities to take seriously the changes and challenges that technology will bring to them in the 21st Century.

The utilization of educational technology up to this point has given rise to the opportunity and potential of restructuring the educational system to extend the learning environment beyond the four walls of the classroom. In order to integrate technology and education we must first develop our view of the revolution that technology is creating in education, and appropriately incorporate it into the system (Chen et al., 2006). This in turn will influence cultural changes in the education system, and technology is also often assumed to be the catalyst of new pedagogical change towards e-learning.

Distance educators have had to redefine their communication skills. They have found that two-way interaction is a critical feature of the educational process. Educators must communicate progress to learners as well as engaging the learners into the e-learning environment.

For Distance Education to be successful, high levels of interaction typically need to be present for learners to have a positive attitudes and greater satisfaction. Two-way interactive technologies (e.g., video, audio, audio graphics, and computer conferencing), while capable of providing two-way interactivity, still depend on user skill to successfully bring about interaction in an instructional context.

The current distance education literature addresses distance interaction from the learner's perspective. Shea, Motiwalla, and Lewis (2001) conducted a study to investigate the status of distance education (DE) at 68 higher education institutions. The number of DE students continues to grow. This tremendous growth can be attributed, in part, to shrinking budgets and lower local student enrollments at universities. DE via the Internet provides them with a low-cost, flexible option to expand into global markets (Casey, 1998). Distance education can provide a strong interaction between the learner, learner/instructor, and the content as well as other learners. There needs to be a distinct interaction with learners and high technology devices. Hence, the four typical types of interaction for distance learners are: (a) learner-content, (b) learner-instructor, (c) learner-learner, and (d) learner-technology. However, distance instructors develop similar, although relatively different, instructional interactions: (a) instructor-learner, (b) instructor-content, and (c) instructor-technology.

These instructional interactions are complex processes; the literature recommends that instructional designers keep them in mind to produce effective, efficient, and quality distance education courses. Transactional distance included the distance that exists in every educational relationship. There has to be a certain amount of structure in distance education that fosters a certain amount of dialogue between the learner and instructor.

On the other hand, Social Presence is a strong communication component that reduces isolation between the distant learner and other learners and instructor. Lack of social presence might affect learner's performance and outcomes during the instructional transaction. Social presence is the degree to which a person is perceived as a real person in a mediated situation (Short, 1976). The notion is that social presence can be transferred both by the medium itself and by the people using the medium for interaction. Transactional distance and social presence are strongly interrelated and together influence the learner's control of the learning process.

NEED OF INFORMATION & COMMUNICATION TECHNOLOGY IN EDUCATION

In philosophy, ontology specifies the most fundamental categories of existence, the elementary substances or structures out of which the world is made. Ontology thus analyzes the most general and abstract concepts or distinctions that underlie every more specific description of any phenomenon in the world, e.g. time, space, matter, process, cause and effect, system. (Heylighen, 1995). It is sometimes confused with epistemology, which is about knowledge and knowing.

In a certain context the term ontology can be confusing. For example, in information technology, ontology can also be seen as a specification of a conceptualization in the context of artificial intelligence and knowledge sharing. This means that an ontology is a description (like a formal specification of a program) of the concepts and relationships that can exist for an agent or a community of agents. This definition is consistent with the usage of ontology as set-of-concept-definitions, but more general. (Gruber 1993)

From the paradigmatic point of view, an ontological definition of the research object is one fundamental aspect which differentiates sciences and branches of sciences. The idea of understanding fundamentally different ontologies in natural science, psychological science and social science was first suggested by Karl Popper (Popper, 1974). According to this tripartition, reality consists of three 'worlds'. World 1, which comprises physical matter, World 2, which comprises the subjective experiences and World 3, which comprises the social facts.

The worlds of Popper were considered different from each other. Modern theory of social systems is based on the idea that the social systems, psychic systems and organic systems are qualitatively different systems, but meaningful experience always consist of them all. In this sense, from our experience point of view, they are more like dimensions of experience than separate from each other. The three systems can be seen to be dependent on each other's existence but they are fundamentally different self-organizing systems. This distinction is constitutive and ontological by nature where ontologically different entities are complementary for each other. In the framework of an ICT based operation environment, technology, learning as an individual practical process and education as a culturally defined institution can be seen to be ontologically different, but complementary. From the social systems theory point of view, these entities can be seen as an environment for each other (Vanderstraeten 2000).

The question in multidisciplinary research is whether there is any shared common ontological assumption about the research target among the different research. The ontology can be similar or complementary in order to form a common ground for the research. The interim understanding based on the context analysis is that there are different ontological assumptions in current research on ICT integration in education but without any integrating or complementary framework. The fact that there are so many different concepts and words defining the phenomenon of ICT integration in education tell us that there may be a different understanding of what they are researching. Not all concepts can be symbolic generalizations of the same phenomenon.

Concepts Referring "Directly" to the Physical World, Technology as Physical Objects

It is quite obvious that in many of the articles there are expressions related to ICT integration in education which are derived from technology itself or refer directly to technology as an object. The only difference in these expressions is the level of complexity. The very concrete expressions refer directly to the physical equipment, like "a PC", "a computer", "a CD-Rom", etc.. The next level is to refer to a computer network, web or Internet, which can still be seen as referring to technology as such, directly without metaphors, but is not concrete any more. A more general expression related to the new operation environment in education is to refer to technology as a general substance by using words like "technology", "ICT". This style is used in the research to maintain relative neutrality and to give an impression of exact and scientific expressions.

In many cases the idea of referring to a physical device like a computer comes from a practical notion that there is new equipment in our physical world which changes the school's or the classroom's physical environment or the environment where the teachers and learner are living. One common concern for the research is what kind of the skills users need to cope with this new environment. According to this type of research, new technical devices are nothing more than advanced and demanding new objects or machines to operate with.

Many different expressions pointing directly to technology appear in the analysed articles. Those expressions that are not really describing ICT integration into education in a micro-context are not related to the topic of this research and therefore are not analysed. Only those expressions referring to the operation environments in education have been noted. For example, *"The Integrated Service Digital Network (ISDN) is a high speed, fully digital communication technology used primarily for interaction of voice, data and image over telephone lines"* (EMI2000-1-1) is not related to the use of ICT in education and therefore is not included in the analysis. *"... use of computers for undergraduate mathematics on the `stand alone' scenario..."* (C&E2001-3-8) is using direct technical expressions and is describing the use of computers in education and therefore included in the analysis.

When referring to ICT integration in education, the research articles are often using quite generic expressions of technology. This could mean that the differentiation of the role and nature of technology in general is poor in the research of ICT integration in education. The other general expressions were using some other references to the technology like ICT, Computer IT, equipment.

In connection with directly referring to technical devices, a "system" metaphor was also extensively used. Especially those documents which were classified as technology oriented contained references to a "system", which basically indicates that the "system" metaphor has been used as a synonym for technology in general. In socially oriented articles and learning oriented articles system reference was not so popular. Of course the number of the references does not tell us anything other than the frequency of expressions; there is still a need to look closer at the micro context of these expressions.

The following chapters will analyse the system expressions more closely and will draw a conclusion on the nature of the metaphorical expressions in the research on ICT integration in education requirements on which great nations are built. It is an investment that takes its time to pay dividends but, according to economists, education generates the highest returns when compared to any other avenues where resources can be committed. In developing countries, where a large population lives at subsistence levels, primary education is a major tool for enriching human capital. At the primary level, young minds are enlightened to accept new ideas, show creativity, develop critical thinking and above all, enable themselves to absorb surrounding information for informed decision-making at any later stage in life. In this regard, computer studies or ICT becomes immensely important. How essential is ICT or computer education at the primary level? This question is worth exploring given that our children are growing up in an information age— and not in an industrial age.

The argument that we are so underdeveloped and devoid of basic infrastructure that we cannot even afford to think about computers is not very valid any more. With decreasing cost and increasing usability of desktop PCs – which in fact face the threat of extinction at the hands of laptops – has turned the situation from bleak to bright. It is now a matter of choice, rather than availability to introduce ICT at primary school level — at least in the urban areas. Here, it is important to note that there are two very different and distinct aspects of ICT in education. One is teaching ICT itself, and the second is using ICT as an augmented tool to the existing teaching methods which is more important, it will be very useful if the people study from MCSE, CCNA, CompTIA, IBM, Ctirix.

Views on the Nature of Information Technology

It seems obvious that a world with information technology is somehow different from a world without information technology. But what is the difference? Is it a difference of order (faster, closer, clearer, etc) or is it a difference of kind? How can we make sense of these questions? Does technology shape society or society shape technology, or both shape each other? What is the nature of this shaping? Is it in practices, in ways of thinking, or is it more fundamental? The answers to these questions will obviously influence the judgments we make about the social and ethical implications of information technology when we consider the policy and practical concerns of using information technology in a particular domain (such as commerce, education or government).

The answer to these questions are also grounded to a large extent in one's particular, implicit or explicit, ontology of information technology itself; what is the nature—way of being—of information technology as such? Obviously many different ontological positions are possible and have emerged. Nevertheless, it may be useful for the purposes of this entry to discern at least three contrasting and prevailing views.

Information Technology as an Artifact or Tool

The most common view of information technology is that it is an artifact or tool simply available for humans to achieve their objective and outcomes. Some of these tools might be useful and others not. When users take up a tool or artifact (word processor, mobile phone, etc) it will tend to have an impact on the way they do things. For example if I write with a word processor I would tend to have different writing practices than I would with pen and paper. According to this view, we need to understand the impact that information technology has on society as it is taken up and used in everyday practices.

For example, how will communication with mobile phones change our social interaction and social relationships? In asking such a question this view does not primarily concern itself with the development of the technology—why and how did it come about in the first instance. It mostly assumes that the particular technology—mobile phones in this case—operates in a more or less uniform manner in different social settings. In other words, it assumes that a particular technology has certain determinate effects on, or in, the context of its use.

This way of conceptualizing information technology leads to questions such as "what is the impact of the internet on education" or "what is the impact of CCTV on privacy". This view of technology is often criticized for a greater or lesser degree of technological determinism. Technological determinism is the view that technology more or less causes certain ways of doing or ways of organizing to come about. For example, a technological determinist may argue that the Internet's open and non-hierarchical architecture can more or less *cause* a society that uses it to become more open and less hierarchical. The work of Postman (1993) is an example of this type of critical evaluation of the impact of technology on society.

Information Technology as Socially Constructed Artifacts and Actors

Many scholars argue that the 'impact view' above of information technology does not give an adequate account of the relationship between information technology and society (Bijker, Pinch, and Hughes 1987, Bijker 1995, Law 1991, Latour 1991). Firstly, it does not take into account that the technology does not simply appear but is the outcome of a complex and socially situated development and design process. In this development and design process many alternative options become excluded in favour of the technology that is now available—obviously with important implications. In other words there are many cultural, political and economic forces that shape the particular options suggested as well as the way the selected options become designed and implemented (Bijker, Pinch, and Hughes 1987).

It is not only technology that 'impacts' on society; technology itself is already the outcome of complex and subtle social processes—in other words it is socially constructed. Moreover, they argue that when we look at the actual uses of particular technologies we discover that users use them in many diverse and often unexpected ways—leading to many and diverse unintended consequences.

Both in its design and in its actual use there is an ongoing reciprocal relationship in which society and technology co-construct each other; they act through and upon each other. It is therefore very difficult to make general statements about the 'impact' of a technology. One can, at most, speak of some general trends for which many exceptions will invariably exist. For the proponents of the this constructivist view it is important to understand, through detailed descriptive accounts, the particular ways in which technologies emerge and become embedded in particular social practices. Examples of such studies can be found in the work of Bijker (1995), Law (1991) and Latour (1991).

5

Tools of Information Technology uses in Education

SYSTEM IMPLEMENTATION STRUCTURE

To implement the WBE we propose the system implementation structure. The school classrooms, office and training centers are connected through Intranet. This Intranet is connected to the Internet by using network operating system. Firewall is introduced between Intranets and Internet in order to provide security.

System Description

In initial stage of school education the students are not expert in reading and writing. Subject understanding increases if they learn the things through visualization. With the help of multimedia or Rich Media, which includes, Audio, video, graphics and Java Applets have made WBE very effective.

In the primary stage students don't have good knowledge of English. Therefore the presentation should be available in their mother tongue for the better understanding, which is also helpful in improving their pronunciation. This is possible by developing Natural Language Interface to database. One important device called, as "Tech Commander" is also useful for teachers to identify students potential by viewing any students computer display on his own monitor. If he finds something that everyone should see he could set everyone's monitor to display it.

Web Based Education: Considerations and Approaches

- *Conversion of Existing Material*: In order to shift from traditional education to WBE we have to convert the existing school educational material to the Web. The important points to consider are bandwidth, design, usability, and the necessity of high quality media elements and consistency of material across the mediums.

- *Authoring for Multiple Delivery Environments*: We have to provide consistency of interface and ease of authoring and design of an effective multi platform course.
- *Using the Web for student/Teacher Interaction*: Web site can be used for posting of assignments, student work and marks, along with the ability to submit work on-line through the site, also JavaScript and JAVA applets to demonstrate course concepts interactively. This means that course delivery on the Web must be dynamic and truly interactive between the instructor and the students.
- *Faculty Support and Training:* We have to provide centralized support and training resources for training the teachers initially.

Problems to be faced while implementing WBE in school education in Indian context:

Looking at how to use Information Technology in school education, its different tools, the system structure as defined and described it is obvious that we will face some problems while implementing WBE in school education in India.

The major problems we will be facing are:

- Intensive Training to Schoolteachers: Schoolteachers are not introduced to the web based education. Therefore training should be given in order to create a learning environment that will itself train and spur students on the one hand to turn the learning experience into useful, practical and personal knowledge.
- WBL awareness and Workshops: In rural area parents are not much knowing about WBL. So the demonstration, seminars and workshops needs be conducted for society in order to understand the importance of it.
- *Bandwidth Limitations*: Limited bandwidth of Internet connection gives slower performance for sound, video and intensive graphics, causing long waits for downloads that can affect the ease of the learning process. Improved bandwidth will help the teacher to solve his problem.
- *Effect on Teachers*: WBL will lead to reduction in manpower as per as teachers are concerned. This will lead to agitations by teacher's organization.
- *Effect on Students*: Although the students will be benefited by WBL there will some part of students opposing this introduction of technology in education.
- Infrastructure: WBE will primarily require free access to Web to all the learners and hence government of India will have to setup nation wide Fibre Optic Cable network.

- *Access*: Every school will not have equal opportunity to information because of access issues. The schools with fewer budgets will always face this problem. This is the major problem as per as India is concerned, as there is big gap between poor and rich communities in India.
- *Download*: The learning material that appears on web needs to be downloaded will require more time. The speed depends on the transmission methods and bandwidth, which is problem as per as India is concerned.

Important Features of Web-Based Learning Environment

While designing Web-Based Learning sites the following important features should be kept in mind:

- *The Online Syllabus*: An online syllabus provides the instructor with a way to change course material easily and as per the requirements in industry, and the student will have a complete and up-to-date picture of the course requirements. Hypertext links to sample relevant disciplinary web sites may be helpful in giving students (and also prospective students) a sense of the disciplinary context for the course.
- *Personal Home Pages*: Personal home pages can be used to foster the sense that the class is not just a collection of isolated individuals but a community of learners who can profit from interacting with one another. Home pages encourage students to learn about each other so as to encourage contact and mutual interests. This helps the learners to create a group with common interest.
- *Interactivity*: Adding discussion forums and chat sessions to your online course is a common way to add an interactive component to a web-based course. There are many implementations of bulletin board and chat session software to choose from. A second method of interactivity is, of course, e-mail. It's a good practice to have an online list of the e-mail addresses of all registered students, the professor, and teaching assistants. This is possible with an e-mail subscription mechanism included in your Online Syllabus.
- *Assignments*: The web page listings of homework assignments, upcoming events and exams can be more interactive than the familiar print counterparts. If some homework assignments, for example, are based on online materials, they can be directly linked to the class schedule. This helps the students to plan the preparations for the examination in systematic way.
- *Announcements*: To be effective, announcements need to be read; for that to happen, students need to know when a new announcement has been posted. Alert sounds or perhaps a blinking link added to

a page can let students know of new announcements, or perhaps, even a mass e-mail to all students in the course. For a home page, or a long life syllabus, various software tools can be used for the subscribers announcement about page changes. All these techniques will attract the learner's attention towards announcements.

- *Testing*: Online drill or practice testing can be used to reinforce material, even if the results are not used as part of a grade. Reading comprehension questions, for example, in short answer or multiple choice formats can provide students with an assessment of their level of understanding of text. This facilitates the students to measure his level of understanding and through continuous assessment he can try to improve his performance.
- *Course Management*: Software should be available to add or delete students from the course, assign user Ids and passwords, create or edit home pages, and manage any open discussion groups. This helps to keep up to date records of students admitted for various courses.
- *Content*: Perhaps the most difficult part of developing a web-based course is creating the online contents. You can begin by transferring your basic lecture materials to the web and integrating media such as sound, images, and video. Remember, to experiment with incorporating some of the new web-based learning paradigms. And finally, come back and rebuild the lecture building its graph structure and using more html facilities.

Other Features of a Web-Based Learning Environment

- Managing cognitive load — the amount of information people can process — is essential to effective teaching or training. Bombarding learners with too much information at once, called cognitive overload, is one of the chief obstacles to learning. This indicates that we should provide only required information in order to avoid cognitive overload.
- Dividing each tutorial lesson into segments (Classroom, Quiz, Lab, etc.) and then further subdividing these segments into a manageable number of chunks allows users to digest new concepts and skills in a manner that prevents overload.
- *Web-based tutorial*: Users will also enjoy a great deal of flexibility in managing their cognitive load, selecting instructional tasks from a menu of lessons, depending upon the amount and kinds of skills they bring with them, and once engaged in a lesson, selecting which portions of that particular lesson they wish to complete. This allows the students to learn the topics in proper sequence and according to his ability of understanding.

- Because the limited capacity of working memory is rapidly overwhelmed by large amounts of new information, frequent opportunities to practice are important. Rehearsal encodes or moves information into long-term memory. The practice assignments can be presented with practice opportunities throughout the classroom portion of the lesson and is also encouraged to complete the practice portion of each lesson. This allows the student to find out how much he has understood at the end of learning a particular module.
- Finally, online testing is used to reinforce material. Elaborative rehearsal involves presenting questions, which allow the user to apply knowledge in an appropriate context, thus encoding it into permanent memory.
- Quiz questions are designed to provide an authentic assessment of user skill levels by calling on the user to apply the appropriate techniques and practices from the lesson.

THE USE OF INFORMATION AND COMMUNICATION TECHNOLOGY

Incorporating dimensions of the metaphor of information ecology in an analysis of the role of the LIS professions in community development helps to identify the close relationship of social equity with ecology and economy in specific contexts. Connecting economy, ecology and social equity as overlapping, interdependent, and mutually encompassing spheres helps LIS professions participate in community development in ways that are more inclusive of people and impacts their lives in more meaningful ways. Historical evidence suggests that efforts to encourage community development are successful only if there is active participation from people in local communities, and they have a significant stake in community building. In the coming years, use of emerging information and communication technology (ICT) can potentially lead towards an increased transfer of authority and responsibility from higher-level agencies to the neighborhood or community.

This transfer will significantly engage and empower people and provide wider benefits to the community's well-being. ICTs can also strengthen the role of the library and information science (LIS) professions in this process of community building for achieving sustainable community development in terms of economy, ecology, and social equity. This paper will discuss different service-oriented examples in which libraries, library practitioners, and information professionals have played a significant role towards ICT-related community development to promote an equitable distribution of society's resources to disadvantaged users. The aim is for community development practitioners to recognize the efforts of LIS professionals and integrate them in community development efforts.

Libraries are viewed as an information ecology that consists of a system of people, practices, values, and technologies in a particular local environment. Such a vision helps to develop our understanding of information not only in terms of the institution or the technology that helps in its creation, organization, dissemination, and access.

The metaphor of information ecology places information in the context of use of the information, in the context of the people who need the information and who are impacted by the institution and its technologies, and in the context of the embedded community and its "messy" realities. This paper interprets the idea of information ecology in the context of a holistic vision of communities. The goal is to identify some potential characteristics of information ecology (in the context of communities) that are important requisites for sustainable community development.

The discussion addresses the expanded function of the LIS professions in community development initiatives to use ICTs to enhance social equity in communities. Incorporating dimensions of the information ecology metaphor in an analysis of the role of the LIS professions in community development helps identify the close relationship of social equity with ecology and economy in specific contexts. In traditional conceptualizations, the notions of economy, ecology, and social equity offer three different domains in which libraries have an impact on communities.

This paper takes a contrasting position where even though the focus is on the social equity component in community development yet the underlying assumption is that social equity, ecology, and economy are intrinsically intertwined. The line of thought emerges from the understanding that LIS professions are valuable partners in building local assets in community development, based on several interrelated factors associated with human, physical, social, financial, and environmental capital to meet community needs.

LIS and Community Development

As community information providers, libraries have a long history of involvement in community development efforts by creating convivial cultures and providing mechanisms for information delivery and community services. Yet, there is tremendous potential for developing these ties further in the future, especially in the context of the "digital divide" and the specific needs of "marginalized" communities. In recent years, based on the tradition that libraries provide democratic access to communities, scholars have recognized that libraries are among the best places for provision of ICT.

As a result, the entire LIS profession is attempting to extend their roles in community development as information providers to disenfranchised populations. This attempt means incorporating cutting edge ICT and advanced computer systems in community development. The first library-based online community information system in the United States, developed by the Pikes Peak Library System in Colorado, began providing dial-in access to community information in 1978. Some of the earliest library-based community networks are still operating, such as the Sacramento Public Library's nearly twenty-year old Community Information Centre. The networks provide a range of facilities and services for community development such as public Internet access, computer-based community information and referral services, online calendar on community events, employment listings, and information about local business resources, amongst others. Another direction of growth of the LIS professions in community development has involved collaborations and partnerships with community networks to make local information accessible to disadvantaged users.

Strategies have incorporated broad-level application of ICT for tasks and applications considered relevant to the everyday lives of lay people, as well as development of training and support infrastructures within particular disenfranchised communities. For example, the overlapping and shared functions between a Free-Net and a public library benefited the community as a whole in the example of The Three Rivers Free-Net (TRFN), a service of the Carnegie Library of Pittsburgh (CLP) that provides Web space, training, in-house equipment, e-mail accounts, and support for developing Websites.

Such collaborations have been especially significant in the context of recent struggles for survival by community networks that face severe economic cut-backs and limited financial support, unreliability of a volunteer-based work force, lack of a permanent social and technical infrastructure, and competing pressures from commercial and business-oriented enterprises. Such collaborations have led the LIS professions to extend their outreach capabilities to provide services to local users on the margins of society. Recent work by leaders in the LIS professions argues for greater recognition of libraries in community development. These leaders note that LIS professions (including libraries) have historically supported "social justice values" and provided service towards "social and community development and the empowerment of groups and individuals".

"Community librarianship" is one area of involvement in which librarians and LIS professionals advocate "a more flexible and diversified form of public library provision which would enable resources to be targeted at the disadvantaged". Community librarianship addresses not only the development of "sophisticated, qualitative research into the complexion and needs of local communities (community profiling), but also the pro-active involvement of the LIS professions in these communities, along the lines of contemporary practice in community work". "Civic librarianship" is a domain that captures the potential role of the LIS professions to support education for a democratic society and strengthen communi-ties and civic society.

Reaching out to under-represented communities, civic librarianship incorporates "participatory community involvement" and civic virtues as a starting point to build "social capital that supports active citizen participation, public problem-solving and deliberative dialogue".

In response to contemporary needs in a networked society, LIS professionals are applying values of community and civic librarianship in new directions. Such efforts involve addressing social equity via the support of ICT use and developing user-centered services for disenfranchised communities.

These initiatives have potential to further ICT-related support for culturally appropriate programmes that reach out to marginalized people in ways that lead to the empowerment of diverse minority and under-represented populations.

An important factor in this vision is the role of the LIS professions in the organization of the "dynamics of technology" to sustain community development so that "individual and collective action be taken to help orchestrate the integration of community resources--human and informational—into a new dynamic continuum of developmental services".

The importance of understanding the human, physical, social, and cultural environment in terms of linkages between social equity, local economies, and the complex realities of communities as ecologies, is essential in this regard.

Communities, Community Development, and LIS

The role of the LIS professions in reaching out to disadvantaged local communities via ICT use. Holistic characteristics of communities are identified that can be supported by the LIS professions to nurture connections between social equity, ecology, and economy and promote sustainable community development. These connections will draw attention to the role of LIS professionals in community building and will apply this information towards fulfillment of other community development initiatives, strategies, and programmes in diverse settings. Components from the metaphor of information ecology in the analysis help us understand the complex interdependent relationships between social equity, economy, and ecology in the different contexts. Broadly, social equity simply means that which is fair and just. However, it is often problematic to arrive at an understanding of what this implies, especially in cross-cultural situations in which value systems and perceptions of rights, obligations, and duties differ. Such a setting "requires that we recognize the values and norms of other peoples" and that our decisions and actions must be "guided by notions of justice and fairness that accept the integrity and validity of other cultures and lifestyles".

This recognition is important for social justice and social equity for minority and under-represented populations in the United States particularly in the context of their use of ICT. As isolated applications, completely disconnected from the realities of holistic experience of the disadvantaged, ICT is not the solution. Understanding ICT in relation to larger social, cultural, economic, political, and other forms of experiences of under-represented users is necessary to make applicable impacts that may be meaningful to them in their everyday lives.

Consequently, the concept of ecology must be applied in the context of communities. An ecological perspective includes recognition of: the interdependency between parts, the relevancy of local initiatives, the situatedness of efforts, and the acknowledgement of local capacity, as well as diversity and empowerment and the promotion of inclusiveness. Economy is an important and intrinsic feature of sustainability in the specific projects.

The metaphor of information ecology helps advocates to see the intersections between the constructs of social equity, ecology, and economy as expressed in the context of community development. "The complete information environment" definition of information ecology as related to communities addresses all of "[the community's] values and beliefs about information (culture); how people actually use information and what they do with it (behaviour and work processes); the pitfalls that can interfere with information sharing (politics); and what information systems are already in place (yes, finally technology)".

Interdependency and Collaboration between the "Parts" of the "Whole"

Bringing in various voices that represent different stakeholders is essential to the success and sustained implementation of community development efforts. Stakeholders include local community organizations and potential users of the ICT to be designed. The underlying assumption is that all who are directly affected "own" the problem, and should be involved in understanding and addressing it. Lessons learned in projects that employed ICT in rural Canada and East Africa recognized the importance of efforts with clear activities, as well as goals and outcomes that are "identified by stakeholders themselves as critical and representative of what can help them be better off".

Multiple stakeholders bring in different perspectives, goals, and interests, and their involvement in defining what they want to accomplish with the technology is crucial to any community development initiative. Disparate voices also form intertwining elements in sustainable communities and these mutually affecting "parts" create a "whole" community that is greater than the sum of its "parts". However, collaboration and consensus-building are difficult, even when emphasis is laid on learning, ongoing adjustment, and transparency to minimize conflict among stakeholders.

In this scenario, the LIS professions can play an important role in community development via the use of ICT. LIS professionals can act as facilitators to establish collaborative partnerships between local stakeholders such as community networks, educational institutions, government and non-government organizations, public and academic libraries, informational centres, and local citizen-based groups and informal social networks. Their role as intermediaries shifts from what San Francisco's librarian Carl Fleishauer calls "building collections" to "making connections".

At the Prairienet Community Network, a nationally recognized computer-based community information network started in 1993 to serve the twin-cities of Urbana-Champaign and the surrounding East-Central Illinois region, strategies involved cross-disciplinary collaborations in support of efforts to promote community development. For instance, a community development initiative undertaken some years ago involved building a collection of Web resources for small business information exchange.

The project entailed building collaborative partnerships between different agencies to develop culturally relevant online resources via participation and feedback from members of the Community Collaboration for Economic Development (CCED). The author was hired as a research assistant to work with the Director of the Network and other CCED members to develop and implement action research strategies.

The strategies included the training of low-income individuals; incorporating feedback from CCED members especially towards evaluation of existing information resources; analysis of business information needs of potential users; development of a collection of web resources to meet those needs; and understanding the social dynamics between CCED and disfranchised members in the community. Such strategies led to making an appropriate "fit" between the provision of small business information services and local expectations of disadvantaged community members.

Understanding the Local: Situated-ness of Efforts

In order to bring positive change via ICT use in community development, efforts need to be "grounded in local settings," a principle central to an information ecology that helps identify what information or a certain technology "means to the people who use it". Local relevance is related to the importance of context in understanding the nature of information and its meanings that are "unique to each case," and embedded in the set of relationships surrounding information exchange.

Adoption and implementation of technology are significantly related to its appropriation in varied settings that are determined by the extent of local control and adaptation that take place. Based on the past experience of community-based technology initiatives, local involvement and steering of policies to the needs of community users are important steps in the process to address broader issues of integration, identity, and power; they also prevent deepening of social exclusion of disadvantaged populations.

The Community Networking Initiative (CNI), a collabora-tion between the Graduate School of Information Science (GSLIS) at the University of Illinois at Urbana-Champaign (UIUC), Prairienet Community Network, and the Urban League, has the following goals:

- Analyzing local information needs and practices;
- Training and delivering computers to teens and adults;
- Providing technical assistance to community-based organizations;
- Enhancing Prairienet Community Network's information content and retrieval capabilities for low-income, African-American neighborhoods.

An important strategy for community development in the CNI project involves the participation of people from within the local disadvantaged communities. This participation has been significant through the acknowledged recognition of local capacity and resources; the contextualized provision of information services to local expectations; and the enhanced contributions of local community members from minority and marginalized groups.

Various challenges were encountered with CNI. For example, while training different groups in the community, a major problem was "how to handle users of different skill levels". It was not possible to develop a "cookie-cut" solution to serve different needs of various participating groups, members of which included advanced adults, novice adults, advanced teens, novice teens, and others. In response to different contextual parameters (shaped by both social and technological factors) such as skill level, comfort with use of technology, social dynamics in each group, and delivery rate of information, a range of strategies were adopted. Strategies varied in terms of the duration of time, learning goals, training mechanisms, phase development trajectories, and broad and specific outcomes.

From these efforts in the application of training procedures and strategies to local populations, an important lesson was learned: the need for flexibility in process development and implementation. A situated understanding of technology recognizes multiple contexts of technology use in different settings. It helped develop a focus on the local context, on people's practices and applications, and on the systems of relations surrounding ICT use.

Understanding the Local

Working with the community and letting its members decide what directions to choose are central for building capacity and engagement in participative democracy for making positive changes for a better quality of life. Empowerment endeavors can systematically support people, organizations, and communities achieve mastery over their own affairs.

These endeavors contribute to creating knowledge and developing culturally appropriate information resources that meet needs of actual users, for such resources are built by members of the community themselves. LIS faculty and students at UIUC are engaged in ongoing collective inquiry into the everyday lives of the community at the Puerto Rican Community Centre (PRCC) located in Chicago's Paseo Boricua neighborhood. The goal is to incorporate localized knowledge and local narratives of community experiences in the online component of the Paseo Boricua Community Library (PBCL). Collaborative practices and equitable interactions with PRCC members are resulting in the development of Community Inquiry Labs (CIL) that include ICT use to "support community learning and action that addresses critical local problems in a democratic manner". CILs include online curriculum units, developed to address intersection of race, class, and technology-related issues that are incorporated into computer training classes for high school students at the Pedro Albizu Campos High School in Paseo Boricua. Training classes engage Latino youth to identify relevant concerns in the community, build culturally relevant community resources, and elicit their feedback about an online community information system that will form a part of the Paseo Boricua online community library.

Measuring Local Capacity: Neighborhood Asset Mapping

Understanding the extent and boundaries of an information ecology in the context of the community calls attention to the measurement of its existing skills and potentials. Community development efforts support community in ways that "assemble its assets and capacities into new combinations, new structures of opportunity, new sources of income and control, and new possibilities of production".

Community capacity is defined as the community's ability to identify, enhance, and mobilize its human promise, economic opportunities, social relationships, and ecological capital for the purpose of improved community stability. It provides a practical way to reach out to "marginalized" populations and target local neighborhoods in order to enhance resident participation and build on existing ICT-related neighborhood assets. LIS leadership in community development can be potent to help develop asset mapping as a community capacity measuring tool, as well as link service initiatives to the local community asset base. Asset mapping is a "visual path displaying capacities and assets" in which" community members identify what is an asset and how it can benefit their community". It recognizes that all kinds of communities, including poor and disadvan-taged user groups, have assets that can be acknowledged within planning efforts.

Grounding strategies to existent geographic areas accommodates an integrative and comprehensive vision of community building. Several campus units at UIUC, such as the Department of Urban and Regional Planning, the Department of Landscape Architecture, the Department of Architecture, and GSLIS, are involved in service-learning activities in East St. Louis where faculty and students work with neighborhood organizations to map community resources and "address the immediate and long-term needs of some of the city's most distressed communities".

Activities involved documentation and mapping of local community assets in order to build geographic information system (GIS) based databases for the visualization of community components and their relationships; to develop urban systems models to analyse census-based economic and environmental variables that impact community develop-ment; and to identify positive and negative factors that impact development of community technology centres across the city.

Diversity and Empowerment

Diversity is an important element of natural information ecology and an inherent feature in contemporary local on-site and online communities. As a Nations Bank official concluded at the end of the conference titled, Blueprint 2000: Accepting and working successfully within an increasingly diverse America is the nation's single largest challenge affecting community building.

If we are afraid of each other, full of hate for each other, or discriminate against one another, how can we expect to build communities that are livable? Within our increased diversity, we must work together to build a civil society. In order to promote such a civil society, it is important for there to be recognition and acceptance of a moral, social, legal, and intellectual equality in "our inherent differences". The participatory action research (PAR) framework gives an opportunity to actualize and operationalize the notion of equality in our practical work and personal lives by empowering all people who interact within settings that adopt the underlying PAR philosophy. PAR has been practiced under different names--action research, collaborative research, community-based research, amongst others.

There are common threads underlying PAR manifestations: a) democratization of the knowledge process in which the people who are usually considered "research subjects" become part of the research process as researchers analyzing their own experiences; and b) emphasis on social justice and social equity via action to change imbalances in distribution of resources, information, and power. Using PAR, participants draw intersections between social equity, economy, and ecology in ways that lead to empowerment of those traditionally "left out" in the process of resource-sharing and decision-making.

Empowerment in PAR comes from letting people on the margins themselves determine what meaningful intersections are and how they should address them in ways to improve their conditions. PAR is significant for community development in the use of ICT since it can be applied towards the process of knowledge creation, as well as in the context of taking action that brings about changes at different personal, social, political, economic, and cultural levels. Such initiatives will lead directly towards the empowerment of participants by helping them recognize their own worth as well as leading to their social recognition as creators of knowledge.

Additionally, PAR will provide an avenue for participants to take action that leads to an improvement in their own experienced situations. The LIS professions can play an important role in PAR processes by helping community development practitioners to reconfigure embedded relationships, practices, and values of people belonging to various segments in the community. One direct outcome of adopting a PAR approach by LIS professionals is the transference of power to make decisions during various stages of project implementation from external sources to the hands of the "insiders" in local communities. The Afya (Swahili for "health") project is an example in which the LIS faculty and students at UIUC are working with community activists to develop participative and democratic practices to bridge the social, economic, and political divides by targeting barriers that prevent access to health information and services experienced by African American women in the community.

Afya involved establishing partnerships with SisterNet, a local network of African American women committed to address their physical, emotional, intellectual, and spiritual health needs. LIS professionals worked towards building collaborations between local SisterNet women and community health care and information providers. Faculty and students at GSLIS helped to facilitate such collaborations, to advance participative techniques, and to develop online and offline resources that supported the work of community groups. SisterNet women shared their realities and experiences, played activist and leadership roles, and took action to bring changes in the community for improving their health-related situations.

Prairienet Community Network was an important partner in the Afya project and provided computer technology, training, and Web development services. Parkland College, the local community college, was a supporter of SisterNet through its Department of Adult and Continuing Education and provided access to computer hardware, software, and other resources. One major challenge in building community connections was the lack of a common vocabulary for communication between traditionally defined technological "experts" and disenfranchised users. Issues were greater than those related to language; barriers were more associated with established mindsets, existing institutional bureaucratic procedures and policies, and individual personality traits.

In such divergently disjoint situations, LIS professionals played significant roles to build bridges across social and technological gaps. Promoting Inclusiveness: Outreach in Planning and Policy Implementation Sustainable information ecology requires inclusiveness in the processes associated with its development, a fact that is also true for ICT use in community development projects. LIS professions can play active roles to promote outreach to local residents from minority and under-represented populations.

The goal is to develop successful user-centreed information services that are effective at the grassroots levels. Towards this vision, greater support in favour of disenfranchised groups is needed from civic and business associations, non-profit and voluntary organizations, public and private schools, community colleges and universities, health centres, and other local institutions. The underlying idea is to work "with" people rather than simply "for" them by integrating scientific knowledge, disciplined effort, and learned skills into a more contextualized interactive practice.

Zimmerman, et al. highlight the examination of grassroots community organizations for promoting empowerment of "marginalized" populations. Participation of local disenfranchised people in voluntary community organizations, for example, is empowering for "marginalized" individuals for it leads to higher competencies, confidence, sense of duty, and lower feelings of helplessness. In the Afya project, participation of SisterNet women insured that efforts were rooted in existing social institutions and activities of community African American women.

Locating computer-training labs in the local Urban League, Workforce Prep Centre, and within Prairienet Community Network's premises was significant in trying to integrate information access strategies within the daily social life of community members. Similarly, strategies such as developing SisterNet Resource Centres for making health-information resources available at local public spaces such as health salons, churches, libraries, and family centres are instrumental in this regard.

In order to incorporate libraries within the daily social experiences of African American women, the Afya project tried to develop public libraries as an important place for promoting collaborations between SisterNet women, health care and information providers, researchers, local organizations, and civic institutions, as well as other community stakeholders. Presenting participative strategies in public libraries, for promoting access to networked digital information for local Black women, was one such initiative. Similarly, the organization of the SisterNet Health Fair in a public library was another initiative that contributed in trust-building and knowledge exchange of relevant health-related social practices and availability of appropriate resources.

Future Directions of Growth

"The modern public library in large measure represents the need of democracy for an enlightened electorate, and its history records its adaptation to changing social requirements". There is tremendous potential for future growth of the LIS professions as gateways of provision and access to the Information Age for all members in the community. In 1990, the American Library Association adopted the policy, "Library Services for the Poor," stating that libraries and LIS professionals should recognize their role "in enabling poor people to participate fully in a democratic society, by utilizing a wide variety of available resources and strategies".

There are strong critics of the LIS professions to achieve the goal of social equity for under-represented populations. Research by scholars like Chatman , Dervin , Bishop et al., and others has shown that libraries are the last places where minority and marginalized individuals seek information. In this context, community development practitioners need to realise the capabilities and capacities as well as some of the following challenges that the LIS professions face in community development efforts: a need for creativity and openness in outreach to disadvantaged community members; progressive strategies to develop increased involvement in local community initiatives; and greater flexibility and simplicity in policies and procedures. To strengthen their role in community development for social equity, LIS professions will have to loosen bureaucratic procedures, reconsider national and local limiting institutional practices, discard rigid administrative frameworks, and redefine policies of libraries and information systems based on incorporating representative community voices in planning and implement-ation deliberations.

Provision of greater proactive service to disenfranchised communities in different geographic areas to promote activities such as community development planning, assertive community asset mapping, and environmental scanning procedures are first steps in that direction. Other steps towards social equity are for LIS professionals to work together with community development practitioners and others and to bring change in broader policy issues via political action and legislation. "Working our collaborative and citizen-based efforts into the formal, local political structure will not only create policy that reflects the values of citizens but will also hasten reform of local government from that of a purely representative form to a highly participatory and dynamic decision-making structure". The goal is to insure social equity for disadvantaged users by initiating whatever it may take, including political, cultural, and/or economic reforms. In order for community development initiatives to be more meaningful to under-represented populations, it is important to recognize the intertwined relationships that occur in real-life situations between social equity, economy, and ecology.

These factors emerge in settings as individual temperaments and personalities respond to social and cultural expectations and real-forces of politics, power, and economics come into play.

Only when the complexities and the "messiness" of everyday life of the "marginalized" are acknowledged can there be actual changes in everyday operations and functioning for lay people.

For only then will community development reflect and represent reality in a more accurate manner; experiences that recognize the constant and multiple intersections between the information needs, daily practices, class and financial realities, and issues of social inequalities in the everyday lives of the disenfranchised.

The metaphor of information ecology as applied in the context of community provides a good mechanism for LIS professionals, community development practitioners, and others to draw connections between social equity, economy, and ecology, as well as to make ICT development and implementation more meaningful in the everyday lives of disadvantaged populations.

Future community development initiatives in LIS require creating constant awareness that even though tools may be technology-driven, the vision is yet to lean towards strengthening community in terms of "the common life of beings who are guided essentially from within, actively, spontaneously, and freely relating themselves to one another, weaving for themselves the complex web of social unity".

As information providers, LIS professionals can play an important role in this regard, not by finding piece-meal and isolated solutions to ICT use for social equity, but instead, by taking a holistic view that recognizes multiple perceptions and brings various elements of communities to work together for its sustainable development. Some helpful steps include recognition of the capacity and strengths of local "marginalized" communities, implementation and use of ICT in relation to larger socioeconomic and sociopolitical experiences of under-represented populations, and provision of ICT functionalities that directly situate themselves in the context of everyday lives of disadvantaged users in terms of what is meaningful to them.

Stronger ties between LIS professionals and community development practitioners can do wonders towards reaching this goal because both emerge from rich service-oriented traditions.

The contemporary realities of ICT inequities and deepening divides in our communities demand greater efforts in collaboration between LIS professionals, community development practitioners, and others for building bridges to support social justice and promote sustainable community-wide development for all segments in society.

Outdoor challenge education has become increasingly popular as a method of developing team cohesion.

Also referred to as adventure learning, experiential learning, and challenge programmes, such outdoor education involves having a group navigate a course of mental and physical challenge activities to improve its team relationships, group dynamics, cooperation, and communication. Such training typically provides a framework for group problem-solving exercises by requiring groups to develop a plan that enables all members to complete the exercise.

The physical elements of the course encourage participants to find strength from within and embrace the support they receive from their fellow team members. While educators often find outdoor activities impossible or impractical to implement, elements of the methods can be adapted for business communication courses and other disciplines in which team building and communication are emphasized. This chapter examines briefly the evolution, purpose, and components of outdoor challenge programmes and describes three indoor activities that incorporate some of the benefits found in such programmes.

Kurt Hahn is generally credited as the first person to use challenging outdoor experience as a medium for the development of the self and of groups. Hahn was approached in 1941 by Lawrence Holt, the head of a large merchant shipping line, to address the problem of the poor survival rate of young sailors which Holt attributed to an inability to rely on their "inner resources."

Hahn developed a programme of physical experiences which were a medium for the sailors to mature and realise their full potential. While Hahn recognized its value, he never advocated adventure as an end in itself, but rather as a training vehicle through which youth would mature.

The key tenets of Hahn's concept may be summarized as an experience which:

- Takes place in the outdoors.

- Is structured to assist individuals to discover and realise their inner resources.
- Is designed to reflect the environment in which the participant is expected to operate.
- Is based on adventure activities which are inherently dangerous and deemed appropriate for meeting the first three objectives.

The establishment of the Outward Bound movement by Hahn popularized the challenge experience to a broad array of participants all over the world. Outward Bound is intended to be catalytic, to encourage change and to help each participant more fully achieve self-knowledge and understanding of others.

An Outward Bound trip, however, usually consists of a five-day wilderness challenge experience , so companies are turning to challenge courses, which are more convenient and shorter. Since the late 1970s, outdoor-based experiential training programmes have become increasingly popular; and surveys indicate that the trend continues.

Present-Day Outdoor Challenge Education

Entwined in the experiential learning process is a creative problem-solving process. The word "creative" differentiates the particular type of problem solving from an analytical or technical problem-solving process, in which equations or formulae may be used to reach a singular, "correct" solution. In creative problem-solving, a number of different solutions may be possible, but the solution sought is one that best fits the complex context of the problem. The "accuracy" or "correctness" of the solution is not the issue. Rather the issue is, "What can we do which will give us a very good chance of solving the problem?"

Challenge activities involve mental and physical challenges requiring strong individual and team skills. A facilitator may lead the group through the activities, stopping between them to discuss how the group worked together. During the debriefing, the group discusses how they decided to solve the problem, who took the lead, who held back, and how they can do better in the future. Even the most powerful training experience fades if it is not linked to the reality of the workplace.

For this reason, most reputable outdoor-training companies integrate the outdoor activities with classroom discussions, needs assessments, and extensive preparation and follow-up.

Outdoor training courses typically consist of "low elements" and "high elements." Low element activities rarely take place above eye level. Equipment can include props Or permanently installed "low ropes" courses. Some popular low element activities are summarized as follows:

- *The Web*: A group faces a spider web of rope strung between two trees. The challenge is to get the entire group through the web - without touching it and using each gap in the web only once. People have to let go and allow themselves to be hoisted into the air and put through the web.
- *The Beam*: Each group member must get over a six-foot-high beam between two trees without aids (other than help from the team). Once a participant is over the beam, he or she cannot help those who are not, except by "spotting" or watching them in case they lose their balance.
- *The Islands*: A shipwrecked group must travel from a large "island," which is sinking, to two successively smaller ones using only the ship's plank (a board). The plank, however, is not quite long enough to serve as a complete bridge between the islands. If participants touch the (pretend) shark-infested waters around the islands, they must start over.
- *Electric Fence*: Each member of the group must move from one side of the "electric" fence to the other, without touching the fence.
- *Tube Pass*: A bicycle inner tube must be passed around a circle of people who are holding hands, without either breaking the circle or letting the tube touch the ground.

High element activities often focus on a series of rope-based activities placed 4 to 40 feet above the ground, rock climbing, or rappelling.

Course participants are held secure with harnesses and a system of ropes and pulleys. A climbing tower or wall may also be used. The goals may be to challenge a person to tackle his or her fears, learn to make decisions, and learn to lead a group. The purpose of the high element activities is to push people out of their comfort zones.

The advantages and disadvantages of outdoor experiential learning vary based on the type of programme selected. Some distinct benefits are associated with simply being outdoors and engaging in physical activity. One benefit is that physical risk forces participants to engage actively in the training. A person walking across a cable or falling from a table into the arms of teammates is likely to pay close attention to the experience. In typical indoor programmes, people are less likely to be forced to engage actively and participate because they tend to remain within their comfort zones. Another benefit to outdoor programmes is that participants often experience real feelings and emotions. They are not role-playing, and there is no place to hide from feeling the experience.

Many of the characteristics of outdoor training that contribute to the learning environment (especially in high-impact programmes) may also be those that create danger, emotional distress, and legal liability. In addition, the Americans with Disabilities Act (ADA) raises questions about equal access and fair treatment of employees based on abilities that are not relevant to their jobs. Outdoor activities that require people to touch each other also raise concerns.

For instance, the Spider Web exercise requires a team to lift its members one at a time and pass them through a "hole" in the "web." Some people absolutely love this activity, but others refuse to participate because of the amount of touching it requires. Some see this touching as an invasion of their privacy; the issue can be especially critical when programmes involve multicultural groups. A third concern with outdoor programmes is the outdoor setting itself. Rain, snow, and extremely hot and cold temperatures may not be conducive to the learning experience.

USE OF AUDIO-VISUAL AIDS IN EDUCATION

By audio-visual aids, we usually mean the most modern or the most recently used of these methods. This is a summary identification of very old methods and very modern instruments, and one should react against it. Visual aids are far older. They correspond to a profound tendency among the immense majority of men: to materialize their thoughts in the form of graphic or sonorous images or to give their thoughts a concrete frame of reference. Plato himself took care to set the scenery of his dialogues, and he used concrete words and concrete comparisons as foundations for his most abstract ideas.

In France, the *Très riches heures du duc de Berry* bring out the importance which 'illustration' can take in a work which would have otherwise sunk into oblivion. Xylographic images preceded the printing press by three-quarters of a century and the first illustrated book by nearly a century. The tremendous success of the 'images of Epinal' in books peddled from door to door in France was only a manifestation of popular taste in a society where illiterates continued to be in a majority and where images went with oral literature.

Films, radio and television, considered as educational instruments, have merely developed – at a rapid rate – alongside older means whose importance remains considerable. Their common denominator lies in their function as aids.

This is not a theoretical conclusion, for it is confirmed by the very attitude of the educator. The educator basically must contribute to the training of the individual with a view to his integration into a given society and teach new ideas, facts and techniques to a specific public. It is thus relatively easy to define the goals at which the educator aims. Achieving these goals is another task which brings him face to face every day with the basic problem of pedagogy – that of transmitting or communicating ideas or information. To solve this problem, the educator resorts to infinitely varied means, among them audiovisual aids. If our purpose, therefore, is to aid the educator, we must then offer him as complete an arsenal as possible of these means. But it is the educator and the educator alone who chooses the means which is best adapted to his subject, his audience and his circumstances. It is thus clear that audio-visual aids cannot be separated from educational materials in general. This tendency towards the use of concrete examples has developed through a complex process. At first, graphic representation was probably only a way to enable man to capture fleeting thoughts and the sole way of transmitting thoughts, compared to oral transmission which was subject to rapid distortion.

The invention of writing, a perfect example of a visual aid at its origin, proceeded from the same necessity. It would be interesting to study, for example, in the light of Mayan writing – of the Codex Troano – how man progressed from the talking image to the letter. We can therefore conclude that 'illustrations' were looked upon at first, at least by the most educated persons, as a minor complement to thought. The entire history of publishing until the end of the eighteenth century confirms this. But, in the twentieth century, powerful means of reproduction, associated with radio, cinema and television, have changed the aspect of the problem. Sound and visual 'illustrations' are no longer mere minor complements to thought but they directly influence the thoughts and the very conduct of millions of individuals. It was therefore inevitable that a desire should spring up to master such a powerful instrument, to discipline it for better or for worse. But this coveted mastery is still rather crude: it is often reduced to the creating of a few conditioned reflexes, satisfying the merchant but not the educator. Certain of these audio-visual aids are both means of education and media for information and propaganda, and it is not always easy to draw the line between what belongs to the educator and what is within the province of information or propaganda.

It is also probable that the child is more affected by the violence of street posters and by the shock techniques of radio and television at home than by the visual aids used in school. Should we conclude then that these means are harmful and should we condemn them? This negative attitude would be most unrealistic. The only possible conclusion is to accept the need for basic research in these fields. It should bear essentially upon psychology and upon the social sciences. We educators have already ventured forth, but timidly, onto this terrain. Systematic establishment of contacts with research workers and specialized institutes is the duty of all those who are responsible at the national level for audio-visual services.

It can be reasonably hoped that this basic research will lead to a better use of audio-visual aids and to more scientific pedagogy based upon them. It is not difficult to observe that their use is continuing and developing outside the pale of any research. As a result, a pragmatic pedagogy is taking shape and not necessarily in contradiction to the results of the most scholarly research. Establishing or stimulating closer collaboration between research workers and educators, stimulating the writing of theses or documents containing the fruits of the work of both, and publishing and distributing the results of this work should also be the common task of pedagogical and audio-visual services.

Until now, the problem of the use of audio-visual aids has been examined from an intellectual angle. It also includes important practical and technical aspects. To tell the truth, techniques cannot and should not be separated from pedagogy. We have seen that audio-visual aids cannot be separated from educational materials as a whole, this conclusion being thrust upon us by the attitude of the user when confronted by these materials. Now this same user – whether a teacher, a professor or an adult educator – does not act any differently when pedagogy and techniques are involved. He can never be purely a pedagogue or purely a technician. It is clear, therefore, that the pedagogy of audio-visual aids cannot be separated arbitrarily from audio-visual techniques. No one can hope to achieve good results unless he is a sensitive pedagogue and a skilled technician. The problem must be solved globally.

Unfortunately, this initiation into techniques is not always carried out in the institutions where future educators are trained. In underdeveloped countries, the lack of qualified personnel is the most frequent obstacle to such an initiation. But it is not the only one because similar shortcomings are often found in more favoured countries. Routine, lack of initiative and administrative delays are the main factors responsible for educational sluggishness. There is no doubt that audio-visual aids produce their best results when they are used in connexion with active teaching methods. Here, the task of educators is to draw the attention of their governments to these methods and to the recommendations of previous seminars concerned with the introduction of an initiation into film and radio techniques into normal schools and similar institutions.

Finally, there are other questions which should be taken up in thorough and specific studies. They are related to the use of radio and television in the teaching of reading and writing and of languages.

A great deal has been said about the 'singular, specific and irreplaceable services' which can be rendered, for example, by 'teaching by radio'. A great many hopes were stirred as a result of statements repeated with such warm conviction that one could have believed them to be dictated by experience.

First of all, we should note the ambiguity of the term 'teaching'. In the context of the statements to which we refer, this term covers both the teaching of subjects such as science and history as well as the teaching of reading and writing. One of the greatest problems which remains to be solved is the liquidation of illiteracy. Following hasty conclusions, a belief has grown that, thanks to radio, illiteracy can be liquidated quickly, easily and cheaply. But what do we know about it objectively? Until now, the number of experiments has been limited. Some of them were frankly admitted failures. Fortunately, we will soon be in possession of an exhaustive report of the results obtained during the best known of these experiments, the one at Colombia. One of our experts has made a global study of these results from which we think that we will be able to derive valuable lessons, if not definitive conclusions.

A few complimentary remarks should be made here. The first concerns the basic difference between teaching notions of history, geography, science, etc., by radio and the teaching of reading and writing. We say teaching by radio because the problem of teaching reading and writing by television is infinitely simpler and the results already acquired are sufficiently convincing. At any rate, it can never be said too often that the global use of audio-visual aids always gives the best results. Opposing radio to television or both to films is a typical example of a false problem. In an educational campaign organized and carried out at a national level, all complementary means must be used if possible. It is also virtually certain that results are proportionate to the means employed in a geometrical, not an arithmetical, progression. In other words, overly strict economy does not pay.

There lies a source of misunderstanding. Too often, it has been believed that making an expert and $10,000 worth of equipment available to a government was enough to solve the problem of audio-visual aids in a given Member State. Audio-visual aids do not have this magic quality.

They require serious study and, first of all, serious thought about the place which they should occupy in a budget. Pedagogical problems always end up by leading into budgetary problems – that is, in the long run, economic, social and political problems. The educator must play his pedagogical role. The tool which is offered to him – and this is the case of audio-visual tools – can multiply the activity of the educator in large proportion. Therefore, the political authorities must be convinced of the necessity of a financial effort which is often considerably large. That is a point of view which often escapes the educator: he must also educate administrators and political authorities. Many mistakes begin here.

So far, we have talked about audio-visual 'aids' and 'means'. Educators obviously consider them from this angle in the best of cases. But will this tremendous development of mass information media which we continue to allow itself to be domesticated?

Let us go to the heart of the matter: must we continue to consider these information media as blind forces whose unleashing – and, as far as the educator is concerned, the unleashing begins where his own control ends – would be an educational and cultural catastrophe? Or, to put things in a less impressive but equally embarrassing way, cannot audio-visual 'means' be allowed to play their role without the help or simply the intervention of the educator? Before issuing a condemnation without any possibility of appeal, perhaps it might be wise to remember that films, radio and television can only be arbitrarily separated from the social, economic and cultural context which gives them their means of existence.

No doubt, the study of these problems takes us a long way from modest film strips, flannelgraphs, and even traditional black-boards which are still a luxury for thousands of schools.

Let there be no misunderstanding. We know the importance of pedagogy in the use of audio-visual aids. We know that the training of good educators – in this case, good users of these didactic means – is a long and difficult matter. We know that we must think about the desperate problem of training teachers. But pedagogy itself is only a means whose end is education. And education, in the long run, is only a contribution – naturally, of capital importance – to the integration of the individual into a given society. It is in this perspective – from their production to their final use – that we must look at audio-visual aids and the various questions which they raise.

INSTRUCTIONAL TECHNIQUE AND TECHNOLOGIES

Problem Based Learning and Inquiry-based learning are active learning educational technologies used to facilitate learning. Technology which includes physical and process applied science can be incorporated into project, problem, inquiry-based learning as they all have a similar educational philosophy.

All three are student centred, ideally involving real-world scenarios in which students are actively engaged in critical thinking activities.

The process that students are encouraged to employ (as long as it is based on empirical research) is considered to be a technology. Classic examples of technologies used by teachers and Educational Technologists include Bloom's Taxonomy and Instructional Design.

Theorists

This is an area where new thinkers are coming to the forefront everyday. Many of the ideas spread from theorists, researchers, and experts through their blogs. Extensive lists of educational bloggers by area of interest are available at Steve Hargadon's "SupportBloggers" site or at the "movingforward" wiki started by Scott McLeod. Many of these blogs are recognized by their peers each year through the edublogger awards. Web 2.0 technologies have led to a huge increase in the amount of information available on this topic and the number of educators formally and informally discussing it. Most listed below have been around for more than a decade, however, and few new thinkers mentioned above are listed here.

- Hall Davidson
- Lawrence Tomei
- Karl Fisch
- Ian Jukes
- Jamie McKenzie
- Scott McLeod
- Alan November
- Seymour Papert
- Will Richardson
- Gary Stager
- John Sweller
- Joyce Kazman Valenza
- David Warlick
- David Marcovitz
- George Siemens.

Benefits

Educational technology is intended to improve education over what it would be without technology. Some of the claimed benefits are listed below:

- Easy-to-access course materials. Instructors can post the course material or important information on a course website, which means students can study at a time and location they prefer and can obtain the study material very quickly.
- Student motivation. Computer-based instruction can give instant feedback to students and explain correct answers. Moreover, computer is patient and non-judgmental, which can give the student motivation to continue learning. According to James Kulik, who studies effectiveness of computers used for instruction, students usually learn more in less time when receiving computer-based instruction and they like classes more and develop more positive attitudes toward computers in computer-based classes.
- Wide participation. Learning material can be used for long distance learning and are accessible.

- Improved student writing. It is convenient for students to edit their written work on word processors, which in turn improves the quality of their writing. According to some studies, the students are better at critiquing and editing written work that is exchanged over a computer network with students they know.
- Subjects made easier to learn. Many different types of educational software are designed and developed to help children or teenagers to learn specific subjects. Examples include pre-school software, computer simulators, and graphics software.
- A structure that is more amenable to measurement and improvement of outcomes.

Criticism

Similarly to learning a new task or trade, there is special training that is involved with adding elements of educational technology to the classroom. Without proper training, teachers and students cannot benefit from devices that will improve the quality of education. Devices are a waste of time and money if teachers are not receiving proper training. Another disadvantage is that if teachers plan on using educational technology, they must come to the classroom prior to start time and set everything up and make sure it is all working.

THE SCOPE OF ONLINE EDUCATION

Studying along with working has become quite a common trend among the young generation these days. Unlike the past education system, where pupil are required to visit their respective colleges and universities and attend the classes, the modern generation depends on online education.

We are grateful to the technological advancements that we are able to reap a long list of benefits in this epoch. Keeping pace with the fast and competitive world, students cannot afford to send two or three years concentrating only on studies. Instead they prefer to do any job and continue with their education. The Scope of Online Education seems ever increasing. Since the advent of this type of education, it has been able to attract a large percentage of student population all over the world.

When it comes to secure the Scope for Online Education, one must consider the fact that online education is not limited to only one country and institution. This is considered an education without boundaries, which allows the students to study their chosen subjects at their ease. The Scope of Online Education is widening since more and more universities are offering distance courses on each and every subject you want to pursue. The students all over the world are resorting to this type of education as they get absolutely all the convenience they can expect to get. The choices offered by universities across the world are worth appreciating.

Leaving behind the traditional method of studying, students are becoming more inclined in adopting convenient methods of going about. Since there is an all round growth in the field of online education, most of the interested students are opting for this. Starting from Online Masters Degree and Bachelor Degree to Online Doctorate Degree and Short-term Diploma & Certificate Courses, the Scope of Online Education is increasing to a great extent.

Although this online education system was initially developed in the European countries, gradually countries of different continents are also keeping pace with that. Breaking the national boundaries, students can look forward to get education for the reputed universities all over the world. Keeping in mind the convenience of the students, majority of educational institutions are launching their websites and offering multiple range of courses to the brilliant and interested.

Owing to the fact that people are becoming more used to internet and they have access to internet, the scope for Online Education gets widened. The willing students who look forward to go for international education can surely count on this. Unlike the initial online education, the students are always able to choose from a variety of subjects which are not always found on all the universities. In the modern times Scope of Online Education has increased to a wide extent. With the growing popularity of online education the facilities offered by the reputed colleges and universities have increased. This online education is opening up the option for the students worldwide and enables the students to interact with a large number of other people related to that subject which you are pursuing.

INFORMATION AND COMMUNICATION TECHNOLOGIES

Information and communication technologies, or ICT, allows users to participate in a rapidly changing world in which work and other activities are increasingly transformed by access to varied and developing technologies.

By this definition, you could almost say ICT is technology's version of economic growth, to satisfy the needs and wants of the community over time. ICT is also a very age friendly mechagnmisom ICT tools can be used to find, explore, analyse, exchange and present information responsibly and without discrimination. ICT can be employed to give users quick access to ideas and experiences from a wide range of people, communities and cultures.

Economic Impacts

In recent decades widespread incorporation of ICTs into many tiers of business, political processes and eructuring of the global economy.

ICTs have increased international interconnectedness and sped up the process of globalization. They have been instrumental in the information revolution, facilitating the transition from industrial economies, driven by the manufacturing sector, to knowledge economies. ICTs, in conjunction with globalization and the information revolution, have reshaped the workforce. By increasing the speed of international communication ICTs have enabled corporations to outsource jobs, both in the manufacturing as well as white collar sectors. While this lowers production costs and as a result the cost of goods, it has also had fundamental and often detrimental impacts on labour conditions.

Outsourcing causes geographic fragmentation of commodity chains, in which production of goods occurs in specialized plants in different locations, often traversing international boundaries. Locations with no or minimal restrictions on wages, compensation and entitlements for workers therefor become economically attractive as sites of production. This can lead to the exploitation of workers in developing countries and undermine the bargaining power of organized labour in developed countries outsourcing causes geographic fragmentation of in which production of goods ocure in speciallixes pants in different occasion often traversoi international boundaries despite the international spread of ict the economic impacts have been gerographically uneven.

They have exacerbated prexistion desparities between developed coutries, which can aford to produce and consume the lates.

Despite the international spread of ICTs the economic impacts have been geographically uneven. They have exacerbated pre-existing disparities between developed countries, which can afford to produce and consume the latest technologies and developing countries, which cannot. This gap is known as the digital divide.

Social Impacts

ICTs have impacted societies on many levels. They have extended the reach of public administration, leading to a centralization of regional management into urban centres. They have led to new forms of employment in innovation and production of ICTs and a demand for highly skilled specialists. However ICTs have also enabled professionals in certain industries to be replaced by unskilled workers, or even made entirely redundant. Proponents of ICTs portray this as a 're-skilling' of the workforce, while to detractors it is a 'de-skilling' process.

The diffusion of ICTs within societies is varied, with some institutions and sections of society having greater access to ICTs then others. These divisions are reflected in the content of ICTs. For example the English language, which is understood by only 10% of the worlds population accounts for approximately 80% of internet content.

Despite these imbalance in power relations, many social justice movements believe ICTs can be used to promote equality and empower marginalized groups. These groups advocate ICTs as a means of providing accessible and affordable information and as a platform for voices that might otherwise go unheard.

Economic Development

ICTs have been identified by many international development institutions as a crucial element in developing the worlds poorest countries, by integrating them into the global economy and by making global markets more accessible. The World Bank has collaborated with the International Finance Corporation to promote access to ICTs, an initiative which it describes as one of its most successful information and communication is most important to each and every organation with out IC can not growth org. In 2006 the United Nations launched an initiative called the Global Alliance for Information and Communication Technologies and Development.

INFORMATION AND COMMUNICATION TECHNOLOGIES FOR DEVELOPMENT

Information and Communication Technologies for Development (ICT4D) is a general term referring to the application of Information and Communication Technologies (ICTs) within the field of socioeconomic development or international development. ICT4D concerns itself with directly applying information technology approaches to poverty reduction.

ICTs can be applied either in the direct sense, wherein their use directly benefits the disadvantaged population, or in an indirect sense, wherein the ICTs assist aid organisations or non-governmental organizations or governments or businesses in order to improve general socioeconomic conditions. In many impoverished regions of the world, legislative and political measures are required to facilitate or enable application of ICTs, especially with respect to monopolistic communications structures and censorship laws.

The concept of ICT4D can be interpreted as dealing with disadvantaged populations anywhere in the world, but is more typically associated with applications in developing countries. The field is becoming recognized as an interdisciplinary research area as can be noted by the growing number of conferences, workshops and publications. Such research have been spurred on in part by the need for scientifically validated benchmarks and results, which can be used to measure the efficacy of current projects. Many international development agencies recognize the importance of ICT4D. For example the World Bank's GICT section has a dedicated team of some 200 staff working on these issues.

A good example of the impact of ICTs on development are African farmers getting better market price information and thus not being impoverished by unfair corps buy-out people. Another example includes mobile telecommunications and radio broadcasting fighting political corruption in Burundi. The dominant terminology used in this field is "ICT4D". Alternatives include ICTD and development informatics.

History

The history of ICT4D can, roughly, be divided into three periods:

- ICT4D 0.0: mid-1950s to late-1990s. During this period (before the creation of the term "ICT4D"), the focus was on computing / data processing for back office applications in large government and private sector organisations in developing countries.
- ICT4D 1.0: late-1990s to late-2000s. The combined advent of the Millennium Development Goals and mainstream usage of the Internet in industrialised countries led to a rapid rise in investment in ICT infrastructure and ICT programmes / projects in developing countries. The most typical application was the telecentre, used to bring information on development issues such as health, education, and agricultural extension into poor communities. More latterly, telecentres might also deliver online or partly-online government services.
- ICT4D 2.0: late-2000s onwards. There is no clear boundary between phase 1.0 and 2.0 but suggestions of moving to a new phase include the change from the telecentre to the mobile phone as the archetypal application; less concern with e-readiness and more interest in the impact of ICTs on development; and more focus on the poor as producers and innovators with ICTs (as opposed to just consumers of ICT-based information).

Projects and Anatomy

ICT4D initiatives and projects may be designed and implemented by international institutions, private companies (e.g. Intel's Classmate), governments (e.g. e-Mexico initiative), non-governmental organizations (e.g. International Institute for Communication and Development), or virtual organizations (e.g. One Laptop per Child).

ICT4D projects address one or more of the following issues:

- Access and *Infrastructure*: providing suitable computer hardware, operating systems, software, and connectivity to the internet. These would include the affordability of software and hardware, the ability to share software (as echoed in the Free Software movement), and the ability to sustainably connect to the internet.

- *Capacity building and training in ICT:* installing, maintaining, and developing hardware and software, digital literacy (technological literacy and informational literacy) and e-Awareness.
- *Digital content and services:* e-services (e-learning, e-health, e-business/ e-commerce, e-Governance/e-Government), including concerns related to local-language solutions in computing, and the Open Access agenda.
- *Regulation of the ICT Sector and digital rights:* Universal Access vs. monopolistic structures, Intellectual Property Rights, privacy, security, and digital identity.
- *Ethics and Social Contexts*
- Environment and Agriculture
- Free and Open Source Software
- Gender and ICT
- Health and Medicine
- Policy and Social Analyses
- Technical Innovation for Development.

Problems

Projects which deploy technologies in underdeveloped areas face well-known problems concerning crime, problems of adjustment to the social context, and also possibly infrastructural problems. Literacy issue is one of the factor why projects fail in rural areas, proper education and training to make the user at least understand how to manipulate the application to get the proper information they needed. Constant follow up with the community to monitor if the project is being used or implemented.

Projects in marginalised rural areas face the most significant hurdles. Since people in marginalised rural areas are at the very bottom of the pyramid, development efforts should make the most difference in this sector. ICTs have the potential to multiply development effects and are thus also meaningful in the rural arena. However introducing ICTs in these areas is also most costly, as the following barriers exist:

- Lack of Infrastructure: no power, no running water, bad roads
- Lack of Health Services: diseases like HIV, TB, malaria are more common.
- Lack of Employment: there are practically no jobs in marginalised rural areas.
- Hunger: hungry users have problems concentrating.
- Illiteracy: Text user interfaces do not work very well, innovative Human Computer Interfaces are required.

- Lack of means to maintain the project: some projects may be left to deteriorate in time because if a component gets broken they are costly to repair and maintain.
- Lack of support from the local government
- Social Contexts: the potential users living in rural marginalised areas often cannot easily see the point of ICTs, because of social context and also because of the impediments of hunger, disease and illiteracy.
- Corruption is one of the factors why it hampers the implementation of the ICT project in rural areas.
- Trainings and seminars must be conducted according the suitable time of the farmers, to make sure that their daily routine for livelihood must be done first.
- Proper applications are not user friendly.

The World Bank runs Information for Development Program (infoDev), whose Rural ICT Toolkit analyses the costs and possible profits involved in such a venture and shows that there is more potential in developing areas than many might assume. The potential for profit arsises from two sources-resource sharing across large numbers of users (specifically, the publication talks about line sharing, but the principle is the same for, e.g. telecentres at which computing / Internet are shared) and remittances (specifically the publication talks about carriers making money from incoming calls, i.e. from urban to rural areas). Remittances are estimated to have a volume of upward of 250 billion USD and websites have been established to take advantage of this fact (e.g. Aryty, Philippines; Mukuru.com, Zimbabwe.

Lessons Learned

What's crucial in making any ICT4D effort successful is effective partnership between four key stakeholders:

- Public sector (governments-from developed nations, developing nations, international bodies, and local governments)
- Private sector (companies belonging to members of the target audience, multi-national organizations wishing to expand their markets to the 4 billion people under US$2/day, pro-poor or social companies)
- Informal sector (NGOs, advocacy groups, think tanks)
- Representation from the target audience

InfoDev have published 6 lessons from an analysis of 17 their pilot programmes. These lessons are backed by a variety of examples as well as a list of recommendations, which should be read by everyone starting an ICT4D project.

- Lesson 1: Involve target groups in project design and monitoring.

- Lesson 2: When choosing the technology for a poverty intervention project, pay particular attention to infrastructure requirements, local availability, training requirements, and technical challenges. Simpler technology often produces better results.
- Lesson 3: Existing technologies—particularly the telephone, radio, and television—can often convey information less expensively, in local languages, and to larger numbers of people than can newer technologies. In some cases, the former can enhance the capacity of the latter.
- Lesson 4: ICT projects that reach out to rural areas might contribute more to the MDGs than projects based in urban areas.
- Lesson 5: Financial sustainability is a challenge for ICT-for-development initiatives.
- Lesson 6: Projects that focus on ICT training should include a job placement component.

Sustainability and Scalability

A growing perspective in the field is also the need to build projects that are sustainable and scalable, rather than focusing on those which must be propped up by huge amounts of external funding and cannot survive for long without it. Sustaining the project's scalability is a huge challenge of ICT for development on how the target user will continue using the platform. Development on ICT is not one shot implementation but rather it is a complex process to undertake continuously and the progress of each project evolves around the pervasive education for adaptability of the technology

Also, many so-called "developing" countries, such as India (or other South Asian countries like Sri Lanka, Pakistan, and Bangladesh, as also nations like Malaysia, China, Indonesia, Brazil, South Africa and many others) have proved their skills in IT (information technology). In this context, unless these skills are tapped adequately to build on ICT4D projects, not only will a lot of potential be wasted, but a key indigenous partner in the growth of this sector would be lost. Also there would be unnecessary negative impact on the balance of payments due to imports in both hardware and software. Currently, the main two perspectives coming out of this sector either emphasize the need for external aid to build infrastructure before projects can touch viability, or the need to develop and build on local talent. Both approaches are, of course, not mutually exclusive.

Critics

As it has grown in popularity, especially in the international development sector, ICT4D has also increasingly come under criticism. For instance, questions have been raised about whether projects that have been implemented at enormous cost are actually designed to be scalable, or whether these projects make enough of an impact to produce noticeable change. In Sri Lankan journalist Nalaka Gunawardene argues that thousands of pilot projects have been seeded without regard to generalisability, scalability, and sustainability, implying that these projects will always require external funding to continue running and that their impact is limited. This sentiment echoes a 2003 report by the World Bank. Further criticism on ICT4D concerns the impact of ICTs on traditional cultures and the so-called cultural imperialism which may be spread with ICTs. For example, young males are tempted to spend their recreational time playing violent computer games.

It is emphasised that local language content and software seem to be good ways to help soften the impact of ICTs in developing areas. Anriette Esterhuysen, an advocate for ICT4D and human rights in South Africa, pointed out that some ICT4D projects often gives more impetus to how ICT can help its beneficiaries economically rather than helping them gain a society where social justice and equal rights prevails. She Believes that sustainable development can only be achieved if there is human rights and people can speak freely.

Another point of criticism against ICT4D is that its projects are seldom environmentally friendly. Beneficiary communities are often given the responsibility to dispose of the toxic electronic scrap when an equipment breaks down beyond repair. Since transporting the equipment to a recycling facility is costly; the equipment is often disposed of improperly, thus contributing to the pollution of the environment.

Technology

ICT4D projects typically try to employ low-cost, low-powered technology that can be sustainable in developing environment. Desktop virtualization and multiseat configurations are probably the most simple and common way to affordable computing as of 2009. ICT4D projects needs to be properly monitored and implemented; the system's design and user interface should be suitable to the target users. ICT4D projects installed without proper coordination with its beneficiary community has a tendency to fall short of its main objectives. For example, the usage of ICT4D projects in farming sector in third world countries, where a majority of the population are considered to be technologically illiterate, projects lay idle and sometimes get damaged or become obsolete.

Furthermore, there should be a line of communication between the project coordinator and the user for immediate response to the query or the difficulty encountered by the user. Addressing properly to the problem will help encourage the user for the interactivity and participation.

Organizations

- APC Association for Progressive Communications
- Computer Aid International Computer Aid International
- Dialog Telekom Dialog Telekom PLC Sri Lanka
- LinkNet Zambia LinkNet Zambia
- PhilRice Philippine Rice Research Institute.

Notable Events

World Summit on the Information Society (WSIS)

A major event for ICT4D was the twin WSIS (WSIS)-lead organisation was the International Telecommunications Union (ITU). The first part of WSIS took place in Geneva, Switzerland in December 2003 (with a large ICT4D exhibition and an ICT4D symposium co-ordinated by infoDev). The second part of WSIS took place in Tunis, Tunisia, in November 2005. One of its chief aims of the WSIS process was to seek solutions to help bridge the so-called "digital divide" separating rich countries from poor countries by spreading access to the Internet in the developing world.

Perspectives on the WSIS are available elsewhere on Wikipedia, and this covers links to civil society, Tunis 2005, US priorities at WSIS, media responses, Tunis conference developments, roles for business and government, digital divide issues, the digital divide and the digital dilemma, common ground, a civil society study on WSIS, and external links.

6

Distance Learning Technologies

Until the advent of telecommunications technologies, distance educators were hard pressed to provide for two-way, real-time interaction, or time-delayed interaction between students and the instructor or among peers. In the correspondence model of distance education, which emphasized learner independence, the main instructional medium was print, and it was usually delivered using the postal service. Interaction between the student and the instructor usually took the form of correspondence of self-assessment exercises that the student completed and sent to the instructor for feedback. Formal group work or collaborative learning was very rare in distance education, even though attempts have been made to facilitate group activities at local study centers. Also, traditionally, distance education courses were designed with a heavy emphasis on learner independence and were usually self-contained. With the development of synchronous (two-way, real-time interactive) technologies, such as audio teleconferencing, audio graphics conferencing, and videoconferencing, it is now possible to link learners and instructors who are geographically separated for realtime interaction. However, the type of interaction that takes place is usually on a one-to-one basis, between one learner and another and between one learner and the instructor at one particular time. These technologies are not very suitable for promoting cooperative learning between groups of learners located at different sites. Also, the synchronous nature of these technologies may not be suitable or convenient for many distance learners.

The asynchronous feature of computer mediated communications, on the other hand, offers an advantage in that the CMC class is open 24 hours a day, 7 days a week, to accommodate the time schedules of distance learners. Although CMC systems may be either synchronous (real time) or asynchronous, it is asynchronous CMC, because of its time-independent feature, that is an important medium for facilitating cooperative group work among distance learners.

Current developments in digital communications and the convergence of telecommunications technologies, exemplified by international standards such as ISDN (Integrated Services Digital Network), make available audio, video, graphic, and data communication through an ordinary telephone line on a desktop workstation. Therefore, as we look at distance learning technologies today and look to the future, it is important to think in terms of integrated telecommunication systems rather than simply video vs. audio vs. data systems. More and more institutions that teach at a distance are moving toward multimedia systems integrating a combination of technologies both synchronous and asynchronous that meets learner needs. Therefore, while in the 1970s 'and 1980s many distance education institutions throughout the world used print as a major delivery medium, by the year 2000 many institutions will probably have adopted telecommunications-based systems for the delivery of distance education. This does not necessarily mean that print will no longer be used in distance education. It is more likely that print will be used as a supplementary medium in most telecommunications-based systems, and better ways of communicating information through print will be investigated and incorporated into the design of study guides and other print-based media.

In order to describe the technologies used in distance education, we have selected "The 4-Square Map of Groupware Options" that was developed by Johansen et al. (1991) which is based on recent research in groupware. This model seemed most suitable to our purpose, because we see distance education moving from highly individualized forms of instruction, as in correspondence education, to formats that encourage teaching students as a group and collaborative learning among peers. The "4-square map of groupware option" model is premised on two basic configurations that teams must cope with as they work: time and place. Teams or groups of people who work together *on a common goal deal with their work in the same place at the same time as in face-to-face meetings, and sometimes they must work apart in different places and at different times, as in the use of asynchronous computer conferencing. They also need to handle two other variations: being in different places at the same time, as in the use of telephones for an audio teleconference, and at the same place at different times, as in workplaces, study centers, or laboratories. Based on these configurations, the 4 square model classifies four types of technologies that support the group process: (1) same time/same place, (2) different time/different place, (3) same time/different place, and (4) same place/different time. These four categories are used for describing technologies that currently support distance teaching and learning.

While we use the 4-square model to discuss the major distance education technologies currently being used, we feel that this model does not lend itself very well to discussing new and future developments in integrated telecommunications. Since these integrated systems incorporate many of the features that we classify separately in the 4-square model, we have decided to describe new and future developments in a separate section titled "Future Directions and Emerging Technologies".

Same Time/Same Place Instruction

Same Time/Same Place group interaction is the most familiar format of face-to-face meetings. Certain objectives in distance education programs can only be met by meeting face-to-face. The British Open University, which teaches entirely at a distance brings students on campus during the summer to participate in laboratory experiments. When course objectives require the careful demonstration, observation, practice and feedback of life threatening procedures such as a surgical procedure, it is important to organize face-to-face meetings. In a face-to-face setting accepted practices are only modified slightly to accommodate electronic media. Basic technologies that facilitate a face-to-face meeting involve an overhead projector, a flip chart, electronic blackboard or a projection system that displays computer screens via a LCD monitor. At the more sophisticated end are desk top workstations for each group member which run on special software that helps the group to brainstorm, generate ideas, rank solutions and vote. Also, a record of the group process can be produced at the conclusion of the groups' activities. IBM's Decision Conference Centre in Bethesda, Maryland employs such sophisticated groupware to facilitate group decision making processes. However, innovative approaches are now being adopted to the design laboratory work at a distance by using technologies, as in the dissection of a fetal pig experiment that was designed by the University of Maine using a combination of two-way interactive television, videotape and group work at sites.

Same Time/Different Place Instruction

There are two kinds of Same Time/Different Place Instruction: 1. a meeting through a telecommunications medium or teleconferencing where participants who are separated by geographic distance can interact with each other simultaneously, and 2. the use of non-interactive media such as open broadcast television and radio to instruct a vast number of students at the same time without the ability for the students to call back and interact with the originators of the program. Teleconferencing can be classified into four separate categories depending on the technologies that they use: audio teleconferencing, audiographics teleconferencing, video teleconferencing and computer conferencing. There are two types of computer conferencing: synchronous computer conferencing when two or more computers are linked at the same time so that participants can interact with each other, and asynchronous computer conferencing when participants interact with each other at a time and place convenient to them. Asynchronous computer conferencing is described under Different Time/Different Place instruction. The four major types of teleconferencing vary in the types of technologies, complexity of use and cost. However, they have several features in common. All of them use a telecommunication channel to mediate the communication process, link individuals or groups of participants at multiple locations, and provide for live, two-way communication or interaction.

One advantage of teleconferencing systems is that they can link a large number of people who are geographically separated. If satellite technology is used for the teleconference, then, there is no limit to the number of sites that can be linked through the combination of several communications satellites. In order to participate in a teleconference, participants usually have to assemble at a specific site in order to use the special equipment that is necessary for a group to participate in the conference.

The only exceptions are audio teleconferences which can link up any individual who has access to a telephone, computer conferences that can link up individuals, their computers and modems at home, or direct broadcast satellites that can deliver information directly to participant's homes. However, if more than two people are present at a participating site then it is necessary for the participants to gather at a location which is equipped with teleconferencing equipment in order to participate in a teleconference.

This may restrict access for some learners. In terms of control, participants will have control over the interaction that takes place in a teleconference only to the extent that the instructional design allows for it. However, if the teleconference is taped for later review, students will have more control in the use of the conference. The unique advantage of teleconferences is that they provide for two-way interaction between the originators and the participants. Teleconferences need to be designed to optimize the interaction that takes place during the conference. Interaction needs to be thought of not only as interaction that occurs during the teleconference but pre-and post conference activities that allow groups to interact. Monson (1978) describes four design components for teleconferences: humanizing, participation, message style and feedback. Humanizing is the process of creating an atmosphere which focuses on the importance of the individual and overcomes distance by generating group rapport. Participation is the process of getting beyond the technology by providing opportunities for the spontaneous interaction between participants. Message style is presenting what is to be said in such a way that it will be received, understood and remembered. Feedback is the process of getting information about the message which helps the instructor and the participants complete the communications loop. Monson (1978) offers excellent guidelines for incorporating these four elements into teleconferencing design. The symbolic characteristics and the interfaces that are unique to each medium are discussed with the description of each technology.

Audio Teleconferencing. Audio teleconferencing or audioconferencing is voice-only communication. Even though it lacks a visual dimension, audio teleconferencing has some major strengths: It uses the regular telephone system, which is readily available and a familiar technology; it can connect a large number of locations for a conference; the conferences can be set up at short notice; and it is relatively inexpensive to use when compared with other technologies.

The interconnection medium for an audio teleconference is usually the telephone, which can incorporate microwave, satellite, fibre optic, or coaxial cable transmission. The conference call between three or more persons at different locations is the simplest type of audio teleconferencing. For multipoint teleconferencing among three or more sites, an audio bridge is required to enable sites to interact clearly. The bridge links the telephone lines together so that parties at each location can hear and talk to each other. Olgren and Parker (1983) observe that there are many system options for audio teleconferencing, but the most common forms are: (1) user-initiated conference calls or ("ad lib" teleconferencing), (2) operator-initiated or dial-up or (dial-out) teleconferencing, (3) dial-in or meet-me teleconferencing, and (4) dedicated audio networks. In order to facilitate group-to-group communication, audio teleconferencing requires the use of some type of amplified telephone equipment with a loudspeaker and microphones. The equipment may be built into the room or may be portable. Audio teleconferencing equipment can be described as simplex, quasi-duplex, or full-duplex, depending on the kind of interactivity and interruptibility of the conference connection.

Olgren and Parker (1983) observe that one should keep in mind that voice communication is the backbone of any teleconferencing system, with the exception of computer conferencing. Sophisticated video or graphics equipment can be added to any audio system. But it is the audio channel that is the primary mode of communication. If the audio is of poor quality, it will have a negative impact on users of even the most sophisticated graphics and video technologies. This is very important to keep in mind, because the evaluation of interactive television systems have shown (Dillon, Gunawardena & Parker, 1992) that the most often cited technical problem in television systems is the poor audio quality. While expensive investments have been made in video and graphics systems, very little attention has been paid to the improvement of audio quality in video and audiographics conferencing systems.

Audio teleconferences can be enhanced by adding a visual component to the conference by mailing ahead of time printed graphics, transparencies, or a videocassette to be used during the conference. Each site must be equipped with an overhead projector and a VCR if such graphical or video support is used.

Audiographics Conferencing. Audiographics systems use ordinary telephone lines for two-way voice communication and the transmission of graphics and written material. Audiographics add a visual element to audio teleconferencing while maintaining the flexibility and economy of using telephone lines. Audio teleconferencing is now combined with written, print, graphics, and still or full-motion video information. Most audiographics systems use two telephone lines, one for audio and one for the transmission of written, graphic, and video information.

Currently, the simplest audiographics system is the addition of a fax machine using a second telephone line to an audio teleconference. Printed information can be exchanged during the conference using the fax machine so that visuals can be shared between sites. As a result of recent developments in computer, digital, and video compression technology, fairly sophisticated computer-based audiographics systems are available in the market. These systems combine voice, data, graphics, and digitized still video to create a powerful communications medium.

The PC-based systems have specially designed communications software that control a scanner: graphics tablet, pen, and key board; and video camera, printer, and a modem. One of the key advantages of an audiographics system is the ability to use the screen-sharing feature of the system. Participants at different sites can use different colored pens to create a graphic on the same screen at the same time.

This feature enables the use of collaborative learning methods that involve learners at the remote locations. Since each site is most often equipped with the same types of equipment, it is possible to originate instruction from any location. The systems allow for a higher degree of interaction than one-way video and two-way audio systems. If the system is equipped with a video camera, it is possible to bring video footage to the class or show three-dimensional objects. High-resolution, full-color still video images can quickly be transmitted through dial-up telephone lines. Some systems have incorporated a keypad device that is used for polling participant's opinions and feedback. When the instructor asks a multiple-choice question, participants can use the keypad to key in their response. A central computer tabulates these responses, and the instructor gets an instantaneous statistical summary of the entire group's responses, as well as how each site responded. This is a good way of soliciting and getting feedback from the participants, so that the instructor can adjust his or her presentation depending on the responses received.

Because audiographics systems use regular telephone lines, they are much more cost effective than full-motion video systems. Participants need to be present at locations equipped with the systems in order to participate in a conference, and this may be inconvenient to some learners.

The systems enable the transmission of audio, graphics, data, and still-video information and create a moderate sense of social presence. 'Me human-interface depends to a large degree on the type of communications software that has been designed for the system. Most graphic systems can be mastered by novices with about I hour's training on the system.

Video Teleconferencing. Video teleconferencing systems transmit voice, graphics, and images of people. They have the advantage of being able to show an image of the speaker, three-dimensional objects, motion, and preproduced video footage. The teleconference can be designed to take advantage of the three symbolic characteristics of the medium: iconic, digital, and analog, where the iconic or the visual properties of the medium which is television's foremost strength can be manipulated to convey a very convincing message. Because of its ability to show the images of people, video teleconferences can create a "social presence" that closely approximates face-to-face interaction. Video teleconferencing systems are fully interactive systems that either allow for two-way video and audio, where the presenters and the audience can see and hear each other, or one-way video and two-way audio, where the audience sees and hears the presenter, and the presenter hears only the audience. During a video teleconference, audio, video, and data signals are transmitted to distant sites using a single combined channel, as in the use of a fiberoptic line or on separate channels. Audio is most often transmitted over a dial-up telephone line. The transmission channel can be analog or digital; signals can be sent via satellite, microwave, fibre optics, or coaxial cable, or a combination of these delivery systems.

The term video teleconferencing has become popular as an ad hoc one-time, special-event conference that usually connects a vast number of sites in order to make the conference cost effective. A video teleconference is usually distinguished from interactive instructional television (ITV), which is generally used to extend the campus classroom and carries programming for a significant length of time, such as a semester. ITV may use the same transmission channels as a video teleconference but is distinguished from video teleconferencing because of its different applications: video teleconferencing, an ad hoc conference, and ITV extending the classroom over a longer period of time.

Video teleconferences can be classified into two broad areas according to the technology used for transmission: full-motion video teleconferencing or compressed (or near-motion) video teleconferencing. Full-motion video teleconferencing uses the normal TV broadcast method or an analog video channel that requires a wideband channel to transmit pictures. The range of frequencies needed to reproduce a high-quality motion TV signal is at least 4.2 million Hz (4.2 MHz). The cost of a full-motion video teleconference is therefore extremely high. In the 1970s, conversion of the analog video signal to a digital bit stream enabled the first significant reductions in video signal bandwidth, making video conferencing less cost prohibitive. Therefore, in compressed-video, full video information is compressed by a piece of technology known as a Codec, in order to send it down the narrower bandwidth of a special telephone line. The compressed video method is cheaper and more flexible than the TV broadcast method.

Full-Motion Video Teleconferencing. Full motion video teleconferencing became popular with the advent of satellite technology. For the past decade, educational developers have provided credit courses via satellite television over networks such as the National Technological University (for graduate engineering course), the Arts & Sciences Teleconferencing Service at Oklahoma State University, the TI-IN Network in Texas (for advanced placement high school courses). Both remote and urban schools and businesses have found these educational services valuable enough for their students and employees to make the investment in satellite hardware and tuition fees. Standard C-or Ku-band satellite TV signals can be received by consumer-level hardware costing well under $2,000. For a producer of educational programming, satellite delivery is still more economical than any other format for point-to-multipoint video transmission. Video compression standards and the introduction of fiber-optic cable infrastructure by many telephone and cable companies promises to make terrestrial line transmission of video much cheaper in the near future.

There are, however, at least two reasons that satellite television will probably remain available and, in fact, increase in the foreseeable future. First, there are still many remote areas of the world, even in North America, where telephone service, if it exists at all, is supported by antiquated technology barely able to provide a usable audio or data signal, let alone carry video. These remote areas simply need to point a relatively inexpensive satellite dish powered by solar panels, batteries, or generators-at the appropriate satellite to receive its signal. Additionally, new higher-powered satellites are making it unnecessary to use today's large unwieldy satellite dishes. The new generation of Ku-band satellite is already offering direct broadcast service (DBS) to European households. These receivers, known as VSATs (or very small aperture terminals), are no larger than I to 3 feet in diameter and currently cost less than $500.

The proliferation of smaller, less-expensive satellite television reception technology, along with the continued launching of new, higher-powered satellites, will ensure a continuing niche for this technology to deliver instructional video and data to even the remotest areas of the world that lack other information infrastructure.

Fibre optics is gaining in popularity as a transmission medium for video teleconferencing. Fibre optics is a transmission technology using an attenuated glass fibre hardly thicker than a human hair, which conducts light from a laser source. A single glass fibre can carry the equivalent of 100 channels of television or 100,000 telephone calls, and even more capacity is possible by encasing many fibers within a cable. Fibre optics offers several advantages: It can carry a tremendous amount of data at high transmission speeds; it does not experience signal degradation over distance as does coaxial cable; and it is a multipurpose system that can transmit video, audio, data, and graphics into the school through a single cable. A single fiber-optic cable can carry over a billion bits per second, enabling several video teleconferences to run simultaneously. Many companies, universities, and states in the United States are building fiber-optic transmission networks to carry voice, data, and video.

Video teleconferencing can also use digital or analog microwave systems or dial-up digital transmission lines. Current developments centre on converging the different transmission channels and using a combination of telecommunications channels, satellites, fibre optics, microwaves, and coaxial cables to deliver full-motion video teleconferencing.

Compressed Video Teleconferencing. Videocompression techniques have greatly reduced the amount of data needed to describe a video picture, and have enabled the video signal to be transmitted at a lower and less expensive data rate. The device used to digitize and compress an analog video signal is called a video codec, short for COder/DEcoder, which is the opposite of a modem (MOdulator/DEModulator). Reduction of transmission rate means trade-offs in picture quality. As the transmission rate is reduced, less data can be sent to describe picture changes. Lower data rates yield less resolution and less ability to handle motion. Therefore, if an image moves quickly, the motion will "streak" or "jerk" on the screen.

Currently most compressed video systems use either T-I or half a T-I channel. In a T-I channel, video is compressed at 1.536 Mbps, which is the digital equivalent of 24 voicegrade lines. Many users of T-I codecs opt for transmission at 768 kbps, which is half a T-I channel. The difference in video quality between transmission at 768 kbps and 1.536 Mbps is slight, but the cost savings are significant. With the proliferation of fiber-optic networks, some private video teleconferencing networks are taking advantage of high-quality 45 Mbps transmission. Digital video compression technology has allowed video teleconferencing to become less cost prohibitive. It is not as cost effective as audio teleconferencing and audiographics teleconferencing, but it may soon compete with more-sophisticated audiographics systems with future developments in video compression technology.

Desktop Video Teleconferencing. Future developments in video teleconferencing will move toward integrated desktop video teleconferencing combining audio, video, and data. A fusion of network, personal computer, and digital video has produced the field of desktop videoconferencing. Saba (1993) observes that several telecommunications companies have introduced integrated systems (voice, video, and data) that reside in a desktop computer and provide two-way synchronous communications with voice, image, and file-transfer and screen-share capabilities. This technology allows users to see each other, speak to each other, transfer application files, and work together on such files at a distance. Most systems do not require advanced digital communications technologies such as ISDN to operate. For those wanting to utilize ISDN, it is possible to purchase an ISDN card, while most systems are now being designed to work with telecommunications standards such as ISDN.

Education can use this technology as a method of presenting class material and forming work groups, even though they may be at a considerable distance from each other. An instructor could conceivably present material to the entire class either "live" or through delivery of an audio file to each student's electronic mail account. Students could then work together in real time if they wished to share information over telephone lines.

In one current example, German officials are making use of desktop videoconferencing to form what has been dubbed a "virtual government." As planing progresses to move offices from Bonn, the current capital, to Berlin, planners meet regularly using on-line workstations rather than traveling to meetings. The results provide faster interaction at a much lower cost. As more technologies begin to dovetail, desktop videoconferencing becomes laptop videoconferencing. The use of cellular telephone technology combined with high-speed laptop modems will make it possible for people to hold meetings and work group sessions whether they are at home, in an office, or on the beach.

Interactive Instructional Television (ITV). Interactive instructional television (ITV) systems usually use a combination of "instructional television fixed service" (ITFS) and point-to-point microwave. They can transmit either two-way video and two-way audio, or one-way video and two-way audio, to several distant locations. The advantage of combining ITFS and microwave is that microwave is a point-to-point system, while ITFS is a point-to-multipoint system. Therefore, large geographical areas can be covered by the combination of the two technologies. Microwave connects one location to another electronically with its point-to-point signals, while ITFS distributes that signal to several receiving stations around a 20-mile radius. In the U.S., several states such as Iowa and Oklahoma support statewide networks that use a combination of ITFS, raicrowave, satellite, fibre optics, and coaxial cable.

In an ITFS and microwave television system, the course delivered over the system originates from a "studio classroom" on the campus. The classroom is specially designed to facilitate the extension of a conventional class through television. The audio feedback permits interaction between the teacher and students at distant locations. If a student viewing the class at a remote location has a question, he or she asks it through a talkback system, and it is heard by both on-campus and off-campus class members. The talkback system uses either the telephone or FM microwave technology, called radio talkback.

Interactive instructional television systems also use satellite, fibre optics, or compressed video to extend the traditional classroom. However, these systems are currently not as cost effective as systems that comprise of ITFS and point-to-point microwave.

Integrated Services Digital Network (ISDN).ISDN is a new international telecommunications standard that offers a future worldwide network capable of transmitting voice, data, video, and graphics in digital form over standard telephone lines or fibre optic cable. ISDN transmits media using digital rather than analog signals. In order to move toward a global network, ISDN promises end-to-end digital connectivity, multiple services over the same transmission path and standard interfaces or conversion facilities for ubiquitous or transparent user access. Saba (1988) points out ISDN's applications for distance education: convergence, multitasking and shared communications. Convergence refers to the convergence on audio, video and data media in an integrated telecommunication system. Instruction is possible through voice, data, graphics, and video images. Multitasking refers to the variety of telecomputing capabilities that are available to the learner through integrated telecommunication systems that are based on minicomputers or microcomputers. Learners can gain access to online databases worldwide, and explore multimedia libraries comprising of digital sound, text and images. The shared communications feature allows the teacher and a group of learners separated by distance to work interactively on the same screen, sharing graphics, text, or data at the same time. Therefore, it is possible to solve a problem together or draw a graphic together even though a group of learners may be at different geographic locations. Currently available audiographics systems and desktop video teleconferencing systems provide for the features that will be available in a more user friendly and cost effective manner with the development of ISDN systems.

Broadcast Television and Radio. Broadcast television and radio fall under the classification of same-time/different-place instruction. The difference between broadcast television and radio and the previously discussed technologies under the same category is that broadcast television and radio do not provide for real-time, two-way interaction between presenters and participants. These media, however, can be used to instruct a vast number of students at the same time, even though the students do not have the ability to call back and clarify a statement or ask a question in real time. Many distance education institutions in developing countries, as well as institutions in developed countries such as the British Open University, use broadcast television and radio extensively to deliver programming to a large number of distant learners.

In the United States, while television-both open broadcast cable and ITV-is the most popular media for delivering distance education, radio remains an underutilized medium (Gunawardena, 1988). It is in the developing countries that radio programming has been produced to either support and supplement print-based materials or to carry the majority of the course content.

In the United States, the most common pattern of open broadcast use for delivering distance education is for an institution to make arrangements with the Public Broadcasting Service (PBS) and/or a commercial television station to distribute the educational programming. One of the limitations of this type of distribution is that educational programming is confined to broadcast schedules predetermined by the broadcasting station, which may not be times convenient for students taking the course.

Bates (1984) observes that broadcasts are ephemeral, cannot be reviewed, are uninterruptable, and are presented at the same pace for all students. A student cannot reflect on an idea or pursue a line of thought during a fast-paced program without losing the thread of the program itself. A student cannot go over the same material several times until it is understood.

Therefore, it is difficult for the learner to integrate or relate broadcast material to other learning. Hence, the need for broadcast programming to be accompanied by support materials in the form of prebroadcast notes and follow-up exercises and activities. Research at the British Open University has indicated that "most students find it impossible to take notes while viewing, and those that do are usually very dissatisfied with their notes" (Bates, 1983, p. 61). Access to a videotape of the broadcast, however, will alleviate these problems by giving the learner control over the medium, with the ability to stop and rewind sections that were not clear.

Despite its ability to reach a large section of the student population, open-broadcast television is a one-way communication medium. It does not provide for interaction,(two-way communication) between the student and the teacher and lacks flexibility and ability to respond to student feedback. Since students cannot question the instructor to clarify problems, and since professional broadcast production "makes the learner dependent on 'responsible' broadcasting" (Bates, 1983, p. 61), this system of distribution can encourage passive acceptance of the instruction. To make the system interactive, open-broadcast distribution requires an added system to provide either an audio or audio-video return circuit.

Cable Television. In the United States, cable television began in remote rural areas, expanded into the suburbs, and has now penetrated into large urban areas. Cable has evolved from a way of improving reception in rural areas to a technology capable of providing many channels and even two-way video communication. Microwave relays have enabled cable operators to pick up signals from television stations too distant to be picked up over the air. Satellite interconnection of cable systems makes possible the importation of programming from virtually any part of the world. Today, cable technology is readily available and reaches a large number of homes and apartment units in the United States.

Where cable can provide access to a large section of the population of a given geographic area, it can be used to distribute distance education. Cable can be used to replay Programming offered over open-broadcast television, usually at more convenient times for the students than open-broadcast schedules, or used as a means of delivering nationally distributed television programs, where terrestrial broadcasting facilities are not available.

Interactive cable in most cases is not two-way video. It is one-way video with telephone feedback from the viewer to the instructor, or a technology that provides viewers with one-way video and one-way audio feedback combined with keypads or polling devices with which they can transmit impulses to a central computer in response to questions posed by the instructor. Student responses, such as "yes," "no," "do not understand," "slow down," etc., are immediately summarized by a central computer for the instructor, and often for the viewing audience, thereby adding an element of interaction to the experience.

DEFINITION OF DISTANCE EDUCATION

Distance education brings together students and teachers separated by geographical distance. Many distance education programs operate online, allowing quality student-teacher interaction through live forums, chat, and e-mail. Since online gives people the flexibility to continue working, while continuing on with their educations.. In this competitive environment advancing your career has become more and more difficult. In that competitive environment online education providing platform to enhance your skills.

Distance education offers a wide range of online accredited degrees, includin g online MBA's, teaching certification, and Associate's, Bachelor's and Masters degrees in everything from a network management degree to online criminal justice degrees.

Why Distance Education?

- Learn on your time. Online education allows you to study when you have time—from early morning to late at night.
- No long commute. Your classroom is as close as your living room or home office.
- Keep your job; take care of your family. You don't have to give up important responsibilities to pursue an online education.
- Learn where you live. No need to relocate in pursuit of your online degree.
- Save money. You'll save on room and board, college tuition, classroom supplies, transportation costs, and more.

BENEFITS OF DISTANCE LEARNING

Distance learning technologies offer a myriad of benefits for education including convenience, flexibility, effectiveness, and efficiency. Even since the concept of distance education evolved it has been known to offer several benefits to students who take up such courses. Especially in a country like India Distance education has several benefits to offer. Some of the advantages of students taking up distance education programs or courses are offered below.

In a country which is geographically as huge as India, it might not be possible to construct schools and colleges all over the country in all areas. This is where distance education can be utilized.

For students who might not have quality schools or colleges offering the course that the student wants to take up, then the student can take up a correspondence distance education course with another university or college which may be located else where. Hence inspite of being at home, the student will be able to take up a course and also study for it through the distance education program.

There are several students in India who have financial constraints and hence due to these reasons can not take up the full time courses which they might other wise have wished to. For such students the benefits of distance education are immense, for they can take up a distance education course with out having to pay too high a fees and while also having the option of being able to work and earn money thereby helping them make a living. Hence the distance education courses have several benefits to offer to students.

Distance Education Programme-Tips to Better Learning

Distance learning programmes are those that can be learnt at and from a distance and therefore the terminology came into being and existence. The idea of a distance learning programme as mentioned before in this very website is for convenience and easier method of studying or deriving information as well as learning, and the key to remember is that the method and form of learning is with and at a distance. The programme of distance learning helps in gaining and deriving information in many more fields than one. In the sense, that since this type of education is easier on a person and or as well as a person's schedule, there can be the pursuit of more than one course in the distance education curriculum and pattern that one can pursue simultaneously.

There are tips to better learning that one can imbibe and entail in the scheme of distance education: Learn to contribute and make time for study everyday at home, in this type of education or in this medium of learning-I.e-the long distance education programme. However easy the subject may seem or tough, there has to be a contribution of time and energy everyday towards study, just as one would do if he or she attended a day scholar programme at a school or college (in the vicinity of a classroom)

Learn to take time and understand properly, whatever one is reading as part of the curriculum. Since there is no formal teaching and no teacher to tutor peruse imparting the education as in regular institutions, one has to pay attention and take aid and whenever conducted, attend the actual classes held in campuses on and off for a sealing of the education deal (in the sense, that, to revise and clear doubts, there are periodic or at times, conducted classes at campuses and assigned centers in some cases, for the same).

Do test examinations and time them, so you can sharpen the time keeping and the skills of writing in an actual examination. The idea is to keep time, to be able to train and practice for the actual examination and most importantly, the idea is to sharpen the student nature in a person's personality or in other words, the key is to get competitive in the whole scenario for examinations and study or learning and this is one important tip to follow.

Convenience

Distance learning technologies can provide convenient locations for both students and instructors. Many of the technologies, such as the Internet and telephone, are easily accessed at home. Others, such as videoconferencing, can be distributed from a single point (such as a university) to multiple remote sites (such as schools). Satellite transmissions can be viewed at specified sites, or the transmissions can be recorded for later viewing at home or school.

Flexibility

Many forms of distance learning provide students the option to participate whenever they wish, on an individualized basis. For example, some students may want to review a podcast in the middle of the night or read their e-mail during early morning hours. In addition, one student may wish to spend 30 minutes reviewing a website, while another spends an hour.

Effectiveness

Not only is distance learning convenient, it is also effective. Several research studies have found that distance learning is equally or more effective than traditional instruction when the method and technologies used are appropriate to the instructional tasks, when there is student-to-student interaction and when there is timely teacher-to-student feedback.

Multi-sensory

One of the benefits of distance learning is that there is a wide variety of materials that can meet everyone's learning preference — at least part of the time. For example, some students learn from visual stimuli, such as video, and others learn best by listening or interacting with a computer program. If distance learning courses are well designed, they will likely offer learners a wide range of choices, thereby providing the optimal combinations of interaction and media.

Interactivity

Contrary to popular opinion, distance learning courses can offer increased interactions with students. In particular, introverted students who are too shy to ask questions in class will often "open up" when provided the opportunity to interact via e-mail or other individualized means. Through the increased interactions, teachers can better meet individual student's needs.

Equity

Educational inequity is a major issue in this and other countries. Rural schools often have less contact with educational trends, fewer qualified teachers, and more need for technology.

Distance learning offers great potential for alleviating these issues.

TYPES OF EVALUATION

Evaluation can be either formative, summative, or a combination of both.

Formative evaluation:

- Is an on-going process to be considered at all stages of instruction.
- Will enable the instructor to improve the course as he/she proceeds.

- Facilitates course and content adaptation.
- Will identify major gaps in the instructional plan or the need for minor adjustments.

Some strategies that educators can use to collect formative data from their distant students include:

- Post cards-provide each student with prestamped and preaddressed postcards. On a weekly basis, have students use the postcards to share their concerns or respond to questions during the last three to five minutes of class.
- Electronic mail-Can be a very effective way for instructors and students to communicate. Another plus, while the instructor is eliciting information about classroom learning, students become familiar with the use of electronic mail, a valuable skill.
- Telephone-Call students often. Ask them open ended questions (e.g., "What snags did you run into on the second writing assignment?") to let students voice their concerns. Follow with probes (e.g., "Then, will you need more information sources?"). Set phone-in office hours but be sure to welcome calls at other times.

Summative evaluation:

- Assesses overall effectiveness of the finished product or course.
- Can be a springboard in developing a revision plan.
- Can be a baseline of information for designing a new plan, program, or course.
- Will not help current students since it is conducted upon course completion.

Some questions that educators may want to ask students when collecting summative data include:

- List five weaknesses of the course.
- List three (or five) strengths of the course.
- If you were teaching the course, what would you do differently?
- Student background information: age, level in school, number of distance delivered courses taken prior to this one.
- What would you recommend to a friend planning to take this course?
- What did you think would be covered in this course but was not?
- Would you recommend this course to a friend? Why or why not?

Evaluation Methods

Within the context of formative and summative evaluation, data may be collected through quantitative and qualitative methods.

Quantitative evaluation:

- Involves asking questions which can be statistically tabulated and analyzed, frequently using a scale, check list, or yes/no responses.
- Limits students to responding to the categories made available to them.
- Needs a large student sample for relevant statistical analyses.

Quantitative methods may be most useful for gathering information on large numbers of respondents for whom more in-depth, personalized approaches are not feasible. However, they do have some significant drawbacks:

- Many distance education courses have relatively small class sizes with students from various backgrounds. These small, stratified populations typically defy relevant statistical analysis.
- Quantitative surveys typically result in a rate of return of under 50 percent. A low rate of return often suggests that only those feeling very positively or negatively about the course responded to the evaluation.
- By definition and design, forced choice surveys offer respondents a limited number of possible response options. Therefore, fresh insights and unique perspectives falling outside the provided response categories go unreported.
- The cumbersome and often tedious nature of quantitative data collection can discourage formative evaluation, and often results in an over-reliance on summative evaluation.
- Statistical analysis often results in an illusion of precision that may be far from reality.

Qualitative evaluation:

- Is typically more subjective.
- Involves gathering a wider range and depth of information.
- Is more difficult to tabulate into neat categories.
- Will be less affected by typical small class size.
- Is a more flexible and dynamic method.
- Is not limited to pre-conceived topic of inquiry.
- Allows for student output of topics.

Can use:

- Open ended questioning — with respondents asked to identify course strengths and weaknesses, suggest changes, explore attitudes towards distance delivery methods, etc..
- Participant observation — with the distance educator observing group dynamics and behaviour while participating in the class as an observer, asking occasional questions, and seeking insights regarding the process of distance education.

- Non-participant observation – with the distance educator observing a course (e.g., an audio conference, interactive television class, etc..) without actually participating or asking questions.
- Content analysis – with the evaluator using predetermined criteria to review course documents including the syllabus and instructional materials as well as student assignments and course-related planning documents.
- Interviews – with a facilitator or specially trained individual collecting evaluative data through one-on-one and small-group interviews with students.

What to Evaluate?

Consider the following areas:

- Use of technology-familiarity, concerns, problems, positive aspects, attitude toward technology.
- Class formats-effectiveness of lecture, discussion, question and answer; quality of questions or problems raised in class; encouragement given students to express themselves.
- Class atmosphere-conduciveness to student learning.
- Quantity and quality of interaction with other students and with instructor.
- Course content-relevancy, adequate body of knowledge, organization.
- Assignments-usefulness, degree of difficulty and time required, timeliness of feedback, readability level of print materials.
- Tests-frequency, relevancy, sufficient review, difficulty, feedback.
- Support services-facilitator, technology, library services, instructor availability.
- Student achievement-adequacy, appropriateness, timeliness, student involvement.
- Student attitude-attendance, assignments submitted, class participation.
- Instructor-contribution as discussion leader, effectiveness, organization, preparation, enthusiasm, openness to student views.

Evaluation Tips

- Check out and adapt already published questionnaires; there's no need to re-invent the wheel.
- Draft and revise questions; change if necessary.
- Make use of follow-up probes:
- Alternate between instruction and interaction.

- Sequence your questions for best effect-go ahead and ask for suggestions for improvement before asking for what is good. This will help convey sincerity for seeking improvements.
- Place open ended questions after quick answer questions. This gives students built-in thinking time.
- On summative evaluation, assure anonymity. This can be accomplished by having all questionnaires sent to a neutral site where they would be removed from their envelopes and forwarded to the instructor without a postmark.
- Establish rapport by being interested and supportive. Withhold judgmental responses.
- Adapt to the student in degree of formality and pace of communication.<
- Use evaluation as a method for understanding teaching and learning.
- Try to get both positive and negative feedback. It is important not only to know what is not working, but also what is working.

DISTANCE EDUCATION AS A GLOBAL MOVEMENT

In Europe and other Western countries, a global concern was beginning to emerge. In a recent report, the 12 members of the European Association of Distance Teaching Universities proposed a European Open University to begin in 1992. This is in direct response to the European Parliament, the Council of Europe, and the European Community. In this report, articles from authors in nine European countries describe the use of media and technology in higher education in Europe and reflect upon the need for providing unified educational access in the form of a European Open University to a culturally diverse population.

Telecommunication networks now circle the globe, linking people from many nations together in novel and exciting ways. As the borders of our global community continue to shrink, we search for new ways to improve communication by providing greater access to information on an international scale. Emerging communication technologies, and telecommunications in particular, provide highly cost-effective solutions to the problems of sharing information and promoting global understanding between people. In today's electronic age, it is predicted that the amount of information produced will increase exponentially every year. Since economic and political power is directly related to access to information, many educators like Takeshi Utsurni, president of GLOSAS (Global Systems Analysis and Simulation) have worked to develop models of the "Global University" and the "Global Lecture Hall" which provide resources allowing less-affluent countries to keep up with advances in global research and education.

In the developing world, since the 1950s, the population has doubled to over 5 billion people, most of whom want to be literate and want greater educational opportunities for themselves and their children. The majority of this expanding population is in Asia, where there are massive problems of poverty, illiteracy, and disease. In most developing countries, such as Bangladesh, distance education offers the promise of a system of information distribution through which new ideas, attitudes, and understanding might begin to ooze through the layers of the disadvantaged environments. Drawing upon the well-known model of the British Open University, countries such as Pakistan, India, and China have combined modem methods of teaching with emerging technologies in order to provide low-cost instruction for basic literacy and job training. Turkey has recently joined those nations involved in large-scale distance learning. Only 12 years old, their distance education program has enrolled almost I million students and is the sixth largest distance education program in the world.

Because of the economies of size and distribution, both industrialized and developing countries have embarked on distance education programs. In the early 1980s, record numbers of students in developing countries have gained access to higher education through distance education programs. In many cases, local experts are not available to develop original programs in the language and culture of the people. For this reason, the majority of educational programs are either used intact from the host country or are superficially translated with very few adaptations to the local culture.

When this is done, the results are often unsuccessful. The cultural values of the program designer become dominant, desirable, and used as the standard. There are many examples of programs from North America, Australia, Great Britain, and Europe that were purchased but never used in Africa and Asia because the material was not relevant in those countries. Because the appropriate design of instructional material is a critical element in its effectiveness, the issue of "who designs what and for whom" is central to any discussion of the economic, political, and cultural dangers that face distance educators using information technologies. There have been a variety of efforts to identify theoretical foundations for the study of distance education. Thus far, there has been little agreement about which theoretical principles are common to the field and even less agreement on how to proceed in conducting programmatic research.

Strategies for Teaching at a Distance

Classroom teachers rely on a number of visual and unobtrusive cues from their students to enhance their delivery of instructional content. A quick glance, for example, reveals who is attentively taking notes, pondering a difficult concept, or preparing to make a comment. The student who is frustrated, confused, tired, or bored is equally evident. The attentive teacher consciously and subconsciously receives and analyses these visual cues and adjusts the course delivery to meet the needs of the class during a particular lesson.

In contrast, the distant teacher has few, if any, visual cues. Those cues that do exist are filtered through technological devices such as video monitors. It is difficult to carry on a stimulating teacher-class discussion when spontaneity is altered by technical requirements and distance.

Without the use of a real-time visual medium such as television, the teacher receives no visual information from the distant sites. The teacher might never really know, for example, if students are asleep, talking among themselves or even in the room.

Separation by distance also affects the general rapport of the class. Living in different communities, geographic regions, or even states deprives the teacher and students of a common community link.

Why Teach at a Distance?

Many teachers feel the opportunities offered by distance education outweigh the obstacles. In fact, instructors often comment that the focused preparation required by distance teaching improves their overall teaching and empathy for their students. The challenges posed by distance education are countered by opportunities to:

- Reach a wider student audience
- Meet the needs of students who are unable to attend on-campus classes
- Involve outside speakers who would otherwise be unavailable
- Link students from different social, cultural, economic, and experiential backgrounds

Improving Planning and Organization

In developing or adapting distance instruction, the core content remains basically unchanged, although its presentation requires new strategies and additional preparation time. Suggestions for planning and organizing a distance delivered course include:

- Begin the course planning process by studying distance education research findings. There are several research summaries available.
- Before developing something new, check and review existing materials for content and presentation ideas.
- Analyse and understand the strengths and weaknesses of the possible delivery systems available to you (e.g., audio, video, data, and print) not only in terms of how they are delivered (e.g., satellite, microwave,

fibre optic cable, etc..), but in terms of learner needs and course requirements before selecting a mix of instructional technology.

- Hands-on training with the technology of delivery is critical for both teacher and students. Consider a pre-class session in which the class meets informally using the delivery technology and learns about the roles and responsibilities of technical support staff.
- At the start of class initiate a frank discussion to set rules, guidelines, and standards. Once procedures have been established, consistently uphold them.
- Make sure each site is properly equipped with functional and accessible equipment. Provide a toll-free "hotline" for reporting and rectifying problems.
- If course materials are sent by mail, make sure they are received well before class begins. To help students keep materials organized, consider binding the syllabus, handouts, and other readings prior to distribution.
- Start off slowly with a manageable number of sites and students. The logistical difficulties of distant teaching increase with each additional site.

Meeting Student Needs

To function effectively, students must quickly become comfortable with the nature of teaching and learning at a distance. Efforts should be made to adapt the delivery system to best motivate and meet the needs of the students, in terms of both content and preferred learning styles. Consider the following strategies for meeting students' needs:

- Assist students in becoming both familiar and comfortable with the delivery technology and prepare them to resolve the technical problems that will arise. Focus on joint problem solving, not placing blame for the occasional technical difficulty.
- Make students aware of and comfortable with new patterns of communication to be used in the course.
- Learn about students' backgrounds and experiences. Discussing the instructor's background and interests is equally important.
- Be sensitive to different communication styles and varied cultural backgrounds. Remember, for example, that students may have different language skills, and that humour is culturally specific and won't be perceived the same way by all.
- Remember that students must take an active role in the distance delivered course by independently taking responsibility for their learning.

- Be aware of students' needs in meeting standard university deadlines, despite the lag time often involved in rural mail delivery.

Use Effective Teaching Skills

For the most part, effective distance teaching requires the enhancement of existing skills, rather than developing new abilities. Pay special attention to the following:

- Realistically assess the amount of content that can be effectively delivered in the course. Because of the logistics involved, presenting content at a distance is usually more time consuming than presenting the same content in a traditional classroom.
- Be aware that student participants will have different learning styles. Some will learn easily in group settings, while others will excel when working independently.
- Diversify and pace course activities and avoid long lectures. Intersperse content presentations with discussions and student-centred exercises.
- Humanize the course by focusing on the students, not the delivery system.
- Consider using a print component to supplement non-print materials.
- Use locally relevant case studies and examples as often as possible to assist students in understanding and applying course content. Typically, the earlier in the course this is done, the better.
- Be concise. Use short, cohesive statements and ask direct questions, realizing that technical linkages might increase the time it takes for students to respond.
- Develop strategies for student reinforcement, review, repetition, and remediation. Towards this end, one-on-one phone discussions and electronic mail communication can be especially effective.
- And finally...relax. Participants will quickly grow comfortable with the process of distance education and the natural rhythm of effective teaching will return.

Improving Interaction and Feedback

Using effective interaction and feedback strategies will enable the instructor to identify and meet individual student needs while providing a forum for suggesting course improvements.

To improve interaction and feedback, consider the following:

- Use pre-class study questions and advance organizers to encourage critical thinking and informed participation on the part of all learners. Realize that it will take time to improve poor communication patterns.

- Early in the course, require students to contact you and interact among themselves via electronic mail, so they become comfortable with the process. Maintaining and sharing electronic journal entries can be very effective toward this end.
- Arrange telephone office hours using a toll-free number. Set evening office hours if most of your students work during the day.
- Integrate a variety of delivery systems for interaction and feedback, including one-on-one and conference calls, fax, E-mail, video, and computer conferencing. When feasible, consider personal visits as well.
- Contact each site (or student) every week if possible, especially early in the course. Take note of students who don't participate during the first session, and contact them individually after class.
- Use pre-stamped and addressed postcards, out-of-class phone conferences, and e-mail for feedback regarding course content, relevancy, pace, delivery problems, and instructional concerns.
- Have students keep a journal of their thoughts and ideas regarding the course content, as well as their individual progress and other concerns. Have students submit journal entries frequently.
- Use an on-site facilitator to stimulate interaction when distant students are hesitant to ask questions or participate. In addition, the facilitator can act as your on-site "eyes and ears".
- Call on individual students to ensure that all participants have ample opportunity to interact. At the same time, politely but firmly discourage individual students or sites from monopolizing class time.
- Make detailed comments on written assignments, referring to additional sources for supplementary information. Return assignments without delay, using fax or electronic mail, if practical.

METHODOLOGICAL DISTANCE LEARNING

As distance learning tensions rise in response to concerns about online pedagogy and policy, we need to understand more from faculty who have crossed some of the first hurdles. Where can one go to look for the early adopters or at least those who are less resistant to incorporating the Web in their teaching? Who are the ones to ask about online teaching practices? While previous research indicates that college instructors too often are not utilizing the most sophisticated technologies and interaction opportunities, nevertheless, faculty members were considered ideal sources for providing information on Web-based teaching policies, experiences, training, and incentives in higher education. In this report, we sampled college instructors who had a history of sharing resources on the Web.

Sampling Procedures

Our sampling of instructors employing the Web for teaching and learning purposes comes from two separate sources. First, we selected a random sample of names from The World Lecture Hall. The WLH is an international site first created in 1994 at the University of Texas at Austin to post college syllabi for courses within a variety of academic disciplines. The developers have received national praise and recognition for offering this service.

When beginning to select that sample, however, we noticed the emergence of another resource for faculty and students in higher education. MERLOT was created in 1997 by the California State University Centre for Distributed Learning. It has since expanded to consortia of other institutions and state systems. MERLOT is now a fast growing and free resource intended as an online community of shared knowledge and ideas. In contrast to the WLH, the MERLOT site was originally designed for sharing a wide variety of online learning materials, including assignments, reviews, and member profiles across many academic disciplines within higher education. The capability for peer instructors to review online learning materials was the key feature that distinguished MERLOT from other online resource sharing sites at the time of this study. Even though the WLH and MERLOT members are not representative of all college faculty members, they provide richer online learning backgrounds and experiences than most other available populations. Over 2000 syllabi reflecting more than 80 disciplines and subdisciplines have been posted to the WLH.

Those posting syllabi to the WLH include faculty from religious studies, sociology, theater and dance, accounting, philosophy, marketing, zoology, history, neuroscience, astronomy, nutrition, anthropology, rhetoric, law, and electrical engineering.

At the time of this study, MERLOT contained over 2000 members representing more than 120 different disciplines. Members of MERLOT include faculty from such disciplines as nursing, teacher education, business information systems, geology, arts, computer science, political science, evolution, and theoretical mathematics. The combined sample population, therefore, included a variety of disciplines, degree programmes, and types and sizes of institutions. It also included a wide range of Web expertise.

All these people, however, either had experience posting syllabi online or posting online profiles, critiques, or learning materials. For some in the sample, however, this may have been just a one-time post or brief comment. While the WLH and MERLOT were perhaps the most well known Web sites for resource sharing within higher education at the time of this study, we were not aware of surveys of college faculty representing either or both of these sites. Our random sample during November and early December 2000 included 415 instructors from MERLOT and 286 from the WLH, or a total of 701 instructors from a wide spectrum of disciplines at both sites.1 From e-mail solicitations to this sample, we collected 222 completed surveys; the vast majority were faculty or administrators with additional college teaching responsibilities. While our 32 per cent response rate was generally lower than direct mail or phone surveys, online survey research suggests that this rate is quite good. However, at this time, no expected response rate for online surveys has been firmly established. Nearly fifty different disciplines and subdisciplines were represented in our final sample. Most responses were received from instructors from across the United States, though around 5 per cent of the respondents came from other countries including Hong Kong, Australia, Canada, and the United Kingdom.

Limitations of the Study

As with most online surveys, the present project had several limitations that may have constrained the results and generalizability of the study:

- There are few available resources for faculty online course-sharing, thereby limiting the selection to two of the more popular sites, the WLH and MERLOT. These two Web sites were possibly not representative of all college faculty members who use the Web in their teaching.
- Since users created these sites over long periods of time, many of the collected online faculty member names and e-mail addresses were outdated, incorrect, or changed, especially those in the World Lecture Hall.
- Many of the faculty respondents here were Web savvy and could be described as early adopters of Web technology, thereby inflating any optimistic results regarding online learning experiences and felt need for additional online collaborative tools compared to college faculty in general.
- Tools for teaching and learning on the Web are constantly changing. As a result, it is difficult to generalize many of the findings of this survey related to the utility of particular Web-based instructional tools.
- The online survey instrument was relatively lengthy, effectively lowering the response rate and perhaps causing some inaccurate or skipped responses.
- This survey report labels respondents as college or post-secondary instructors, even though a few of the respondents were in administrative positions with only part-time faculty or teaching responsibilities.

- In an effort to keep the survey at a manageable length, the online survey failed to address key issues such as how courseware tools are funded, the per cent of respondents with tenure, the perceived quality of online certificates or institutes, the forms of online training for instructors, the types of technical support provided for students and faculty working online, how costs are determined for online courses, and perceived learning and motivational factors in online learning. It is hoped that future studies will address such issues.

Despite these limitations, the response rate for this online survey was higher than expected for an e-mail solicited Web survey. In fact, only 7 per cent of those solicited in this particular survey explicitly refused to participate.

DIFFERENT TIME/SAME PLACE INSTRUCTION

This type of instruction usually takes place in a lab or study centre where distance learners gather at different times to interact with instructors, tutors, and other students. Certain types of instructional objectives can only be successfully met by arranging for learners to conduct an experiment in a lab and observing this experiment for evaluation purposes. Local study centers are used by major distance teaching universities such as the British Open University to support the distance learner by offering meetings with tutors, discussion with peer groups, and library facilities. A survey of distance teaching institutions in the United States (Gunawardena, 1988) found that only 41% of the total number of institutions surveyed used local study centers. The types of services provided by most of the institutions were student access to media equipment such as videocassette players and microcomputers, and library facilities such as books, tapes, and cassettes, rather than arrangements for tutor-student interaction.

Different Time/Different Place Instruction

The technologies used in this category are further classified as those that transmit one-way information such as print, audio-and videocassettes, and those that provide for interaction. Technologies that provide for interaction are divided into two groups: (1) those that permit interaction between the instructor and the learner, and among groups of learners such as computer-mediated communication those that provide learner-machine interaction as in computer-assisted instruction (CAI)/computer-based training (CBT) and interactive video and videotex. CAI/CBT, interactive video., and videotex are highly individualized learning experiences that can be designed to give learners control over their learning. Since the technologies that provide learner-machine interaction are discussed elsewhere in this book, they will not be discussed in this chapter.

Print. Until the beginning of the 1970s and the advent of two-way telecommunications technologies, print and the mail system were the predominant delivery medium for distance education. Correspondence study relied primarily on print. to mediate the communication between the instructor and the learner. Currently, many distance education institutions in developing countries use print-based correspondence study as the main distance education medium, as the use of communications technologies is often cost prohibitive. Garrison (1990) refers to print-based correspondence study as the first generation of distance education technology. It is characterized by the mass production of educational materials, and Peters (1983) describes it as an industrial form of education. The difficulty with correspondence education has been the infrequent and inefficient form of communication between the instructor and the students. Also, it was difficult to arrange for peer interaction in correspondence-based distance education. The development of broadcast technologies and two-way interactive media have mitigated the limitations of correspondence study, especially in relation to facilitating two-way communication. However, print remains a very important support medium for electronically delivered distance education. Printed study guides have become a very important component of electronic distance education. In a survey of distance teaching institutions in the United States that use television as a main delivery medium, Gunawardena (1988) found that a majority of institutions cited the study guide, which provides printed lesson materials and guidelines for studying, the most important form of support for distance learners. A study guide can steer and facilitate the study of correspondence texts, television programs, and other components in a distance education course. A study guide, if well designed, can provide the integration between various media components and activate students to read and or listen to presentations of various kinds, to compare and criticize them, and to try to come to conclusions of their own. In a study guide or correspondence text, simulated conversation can be brought about by the use of a conversational tone, advance organizers, mathemagenic devices such as directions, and underlining, self-assessment, and self remediation exercises.

Audiocassettes. Audiocassettes afford the learner control over the learning material, because learners can stop, rewind, and fast-forward the tape. They offer great flexibility in the way they can be used, either at home or while driving a car. Since audiocassettes are a fairly cost-effective medium, they are easily accessible to students. Audiocassettes can be used to tape lectures or can be specially designed with clear stopping points in order to supplement print or video material. For example, in order to facilitate student learning, audiocassettes can be used to describe diagrams and abstract concepts that students encounter in texts. An audiocassette can be used to record the sound portion of a television program if a videocassette recorder is not available, and an audiocassette can provide a review of a television program in order to assist students to analyse the video material. The audiocassette can also be used to provide feedback to student assignments and is a very useful medium to check student pronunciation when languages are being taught at a distance. Audiocassettes can be an excellent supplementary medium to enrich print or other media and can provide resource material to distance learners. Since they can be produced and distributed without much cost, audiocassettes are also a very cost-effective medium for use in distance education.

Videocassettes. Videocassettes are like broadcast television in that they combine moving pictures and sound, but unlike broadcast television, videocassettes are distributed differently and viewed in different ways. An institution using videocassettes for distribution of video material to distant learners can use them as (a) a copy technology for open-broadcast, satellite, or cablecast programming; (b) a supplementary medium, for instance, providing the visual component for educational material carried over audio teleconferencing networks; (c) a specially designed video program that takes advantage of the cassette medium, e.g., its stop/review functions, so that students can be directed at the end of sequences to stop and take notes on, or discuss, what they have seen and heard.

An important advantage in using videocassettes is that students can exercise "control" over the programming by using the stop, rewind, replay, and fast-forward features to proceed at their own pace. Videocassettes are also a very flexible medium allowing students to use the cassettes at a time that is suitable to them. Bates (1987) observed that the "videocassette is to the broadcast what the book is to the lecture".

If videocassettes are designed to take advantage of their "control" characteristics and students are encouraged to use the "control" characteristics, then there is opportunity for students to interact with the lesson material. Students can repeat the material until they gain mastery of it by reflecting on and analyzing it. The control features that videocassettes afford the learner give course designers the ability to integrate video material more closely with other learning materials, so that learners can move between lesson material supplied by different media. "The ability to create 'chunks' of learning material, or to edit and reconstruct video material, can help develop a more-questioning approach to the presentation of video material. Recorded television therefore considerably increases the control of the learner (and the teacher) over the way video material can be used for learning purposes".

Bates (1987) discusses the implications of the "control" characteristics for program design on videocassettes: (a) use of segments, (b) clear stopping points, (c) use of activities, (d) indexing, (e) close integration with other media (e.g., text, discussion), and (f) concentration on audiovisual aspects. When videocassettes are used in a tutored video instruction (TVI) program, where tutors attend video-playback sessions at workplaces or study centers to answer questions and to encourage student discussion, students can take advantage of the features of a lecture (on videocassette) and a small-group discussion, which gives them the opportunity for personal interaction available in on-campus instruction.

Computer-Mediated Communication (CMC). CMC supports three types of on-line services: electronic mail (e-mail), computer conferencing, and on-line databases. In e-mail systems, a message is routed by the system to the addressee's mailbox on the host computer and remains there until it is read by the addressee. This message can be read, replied to, left in the mailbox for later perusal, saved to the hard disk on the microcomputer, deleted, or forwarded to someone else. Most e-mail systems have a bulletin board feature that allows users to read and post messages and documents to be seen by all. However, the messages in the bulletin board system are not linked to each other and provide for only a very limited form of group communication.

Computer conferencing systems, on the other hand, provide a conferencing feature in addition to e-mail, which supports group and many-to-many communication. In these systems, messages are linked to form chains of communication, and these messages are stored on the host computer until an individual logs on to read and reply to messages. Most conferencing systems offer a range of facilities for enhancing group communication and information retrieval. These include directories of users and conferences, conference management tools, search facilities, polling options, cooperative authoring, the ability to customize the system with special commands for particular groups, and access to databases. Databases can be made available on the same host computer used for an e-mail or computer conferencing system, or users can access public or private databases resident on other computers. Some of the well-known computer conferencing systems are: EIES, PARTI, CAUCUS, CONFER, COSY, VAX NOTES, and TEAMATE. Recent developments in groupware, the design of software that facilitates group processes especially in the CMC environment, will have a tremendous impact on facilitating group work between participants who are separated in time and place.

The key features of computer conferencing systems that have an impact on distance education are the ability to support many-to-many interactive communication and the asynchronous (time-independent) and place-independent features. It offers the flexibility of assembling groups at times and places convenient to participants. The disadvantage, however, is that since on-line groups depend on text-based communication, they lack the benefit of nonverbal cues at facilitate interaction in a face-to-face meeting. Levinson (1990) notes that research into education via computer conferencing must be sensitive to the ways in which subtle differences in the technology can impact the social educational environment. "The importance of social factors suggests that 'computer conferencing' may be a better name for the process than is 'computer-mediated communication'; the term 'conferencing' accentuates the inherent 'groupness' of this educational medium" (p. 7). Harasim (1989) emphasizes the necessity to approach on-line education as a distinct and unique domain. "The group nature of computer conferencing may be the most fundamental or critical component underpinning theory building and the design and implementation of on-line educational activities". Gunawardena (1993) reviews research related to the essentially group or socially interactive nature of computer conferences, focusing on factors that impact collaborative learning and group dynamics.

Globaled, a project that linked graduate classes in six universities-San Diego State University, Texas A&M University, University of New Mexico, University of Oklahoma, University of Wisconsin-Madison, and the University of Wyoming-to engage in the discussion of research related to distance education, is an example of the potential of computer conferencing to link students and instructors in learning communities (Gunawardena, Campbell Gibson, Cochenour, Dean, Dillon, Hessmiller et a]., 1994). While the six major participating universities conducted research projects and moderated the discussions of their findings on Globaled, several interested students and faculty from other U.S. and overseas universities, including the Pennsylvania State University and the University of Wollongong in Australia, participated in the discussions. The Globaled community had approximately 90 participants. Globaled was premised on a learner-centred collaborative learning model in which the learner would be an active participant in the learning process involved in constructing knowledge through a process of interaction and discussion with learning peers and instructors.

EFFECTIVE DISTANCE EDUCATION

Without exception, effective distance education programs begin with careful planning and a focused understanding of course requirements and student needs. Appropriate technology can only be selected once these elements are understood in detail. There is no mystery to the way effective distance education programs develop. They don't happen spontaneously; they evolve through the hard work and dedicated efforts of many individuals and organizations. In fact, successful distance education programs rely on the consistent and integrated efforts of students, faculty, facilitators, support staff, and administrators.

Key Players in Distance Education

The following briefly describes the roles of these key players in the distance education enterprise and the challenges they face.

Students -Meeting the instructional needs of students is the cornerstone of every effective distance education program, and the test by which all efforts in the field are judged. Regardless of the educational context, the primary role of the student is to learn. This is a daunting task under the best of circumstances, requiring motivation, planning, and an ability to analyse and apply the instructional content being taught. When instruction is delivered at a distance, additional challenges result because students are often separated from others sharing their backgrounds and interests, have few if any opportunities to interact with teachers outside of class, and must rely on technical linkages to bridge the gap separating class participants.

Faculty -The success of any distance education effort rests squarely on the shoulders of the faculty. In a traditional classroom setting, the instructor's responsibility includes assembling course content and developing an understanding of student needs. Special challenges confront those teaching at a distance. For example, the instructor must:

- Develop an understanding of the characteristics and needs of distant students with little first-hand experience and limited, if any, face-to-face contact.
- Adapt teaching styles taking into consideration the needs and expectations of multiple, often diverse, audiences.
- Develop a working understanding of delivery technology, while remaining focused on their teaching role.
- Function effectively as a skilled facilitator as well as content provider.

Facilitators -The instructor often finds it beneficial to rely on a site facilitator to act as a bridge between the students and the instructor. To be effective, a facilitator must understand the students being served and the instructor's expectations. Most importantly, the facilitator must be willing to follow the directive established by the teacher. Where budget and logistics permit, the role of on-site facilitators has increased even in classes in which they have little, if any, content expertise. At a minimum, they set up equipment, collect assignments, proctor tests, and act as the instructor's on-site eyes and ears.

Support Staff -These individuals are the silent heroes of the distance education enterprise and ensure that the myriad details required for program success are dealt with effectively. Most successful distance education programs consolidate support service functions to include student registration, materials duplication and distribution, textbook ordering, securing of copyright clearances, facilities scheduling, processing grade reports, managing technical resources, etc.. Support personnel are truly the glue that keeps the distance education effort together and on track.

Administrators -Although administrators are typically influential in planning an institution's distance education program, they often lose contact or relinquish control to technical managers once the program is operational. Effective distance education administrators are more than idea people. They are consensus builders, decision makers, and referees. They work closely with technical and support service personnel, ensuring that technological resources are effectively deployed to further the institution's academic mission. Most importantly, they maintain an academic focus, realizing that meeting the instructional needs of distant students is their ultimate responsibility.

INTRODUCTION TO E-LEARNING

e-Learning is the convergence of the web and learning on all levels, whether it be elementary school, college, or business. Knowledge is now considered a competitive advantage and a company's most important asset. Many facts, figures, and forces behind e-Learning are not only causing excitement in boardrooms across the world but are also making Wall Street and the investment community take a closer look.

e-Learning is made up of several methods of learning, which are enhanced or facilitated by technology. As a component of e-Learning, web-based or online learning is likely to be the fastest-growing method for delivering education and training.

Megatrends in areas such as demographics, technology, globalization, branding, consolidation/privatization, and outsourcing will greatly affect the way we learn. These megatrends will affect all learning markets including early education, K-12 education, post-secondary education, corporate training, and consumer products and services. The dearth of skilled labor in the corporate world will only increase the need for e-Learning.

One of the biggest trends affecting the size of the e-Learning market is the astounding growth on the Internet. In the next three years, devices on the Internet are expected to grow from 147.4 million to 345.6 million. In the four-year period between 1998 and 2002, 223 million new users will be online worldwide. By 2002, 21.9 million more kids and 16.6 million more teens will be on the Web. Global online ad spending will reach $33 billion by 2004 and e-Commerce is expected to top the $1 trillion mark by 2003.

Many giants in the technology world are investing in and providing advanced products for and services tailored to the learning market.

These companies, which are expected to make great inroads into e-Learning, include AOL, Yahoo, Microsoft, IBM, AT&T, Sun Microsystems,

Oracle, and Harcourt. The global education and training market is a $2 trillion industry, with $740 billion in the U.S. Approximately 10% of the $740 billion is "for-profit" business.

The growth rate for the different education and training markets is projected at 10-15% but, at least in the corporate sector, e-Learning will far outstrip classroom training over the next few years, claiming almost half of the overall corporate training market.

Educational trends include home schooling, which is growing at 15% ($1.3 billion market). Ten percent of students are attending private schools ($2.6 billion market). The number of high school graduates is expected to increase 20% by 2008, while working adults represent nearly half of all post-secondary students.

Educational software was a $7 billion market in 1997 and is expected to grow by 14% over the next few years; $2.5 billion of the $6.1 billion supplemental educational materials market is sold to consumers, and educational toys are the fastest growing segment of the toy market. All these trends seem to reflect both demographic and societal movements toward providing a better education for anyone who wants to learn.

In the lifelong learner market, self-help books grew at a 21% rate from 1993-1997, and 87% of people that go online do so to pursue information about a hobby or lifelong interest.

By 2003, corporate training will expand to $62.5 billion with $15 billion in outsourced training. Currently, small-to medium-sized businesses are being overlooked in the emerging e-Learning space but probably not for long. Nearly 60% of the workforce is employed by small-to medium-sized businesses and employees of these smaller businesses are more likely to turn to the web for training resources.

E-LEARNING—CONCEPT & NATURE

E-learning is the unifying term to describe the fields of online learning, web-based training, and technology-delivered instruction.

Although we have strong opinions about where eLearning is likely to go and how quickly it may or may not get there, this page is dedicated to linking you to the myriad of e-learning resources in the e-world.

1. A definition of e-learning
2. Websites focused on e-learning issues
3. Organizations you might want to know about
4. Good articles and magazines
5. Interesting books

Definition of e-learning

e-Learning. eLearning. "e"learning. However you write it, definitions abound. The convergence of the Internet and learning, or Internet-enabled learning. The use of network technologies to create, foster, deliver, and facilitate learning, anytime and anywhere. The delivery of individualized, comprehensive, dynamic learning content in real time, aiding the development of communities of knowledge, linking learners and practitioners with experts.

A phenomenon delivering accountability, accessibility, and opportunity to allow people and organizations to keep up with the rapid changes that define the Internet world. A force that gives people and organizations the competitive edge to allow them to keep ahead of the rapidly changing global economy. With good design and delivery, e-learning does all these things. But, at its heart, it is, simply, learning. Too bad most interpretations focus on the technology (the "e") and not on the learning. e-Learning has to keep the people it's designed for in mind. How do we learn? How do we acquire and retain skills and information to help us develop? Only when we address individual learning styles can the "e" in e-learning factor in. Then the technical side—the electronic delivery—can be adapted to the learner. Maybe the "e" should actually *follow* the word "learning." Want more? E-Learners.com explains what is e-learning, how does it work, and some of the terms you'll find when you take e-learning courses. And more still? What is e-learning? (from HowStuffWorks.com) explains what it is and how it works.

"The effectiveness of an elearners experience is greatly enhanced through student-centred (usability) design. For example, students remember more information from a text book that is well organized, with extensive visuals, reflection/interaction points, clear headings, etc. The same concepts exist for online courses—learners learn better through use of clear headings, limited distracters, visuals, screen-friendly fonts, appropriate white space, web safe colors, etc. Basically, usability is the process of testing (through observation) how students behave with a course—what works, what doesn't, what confuses."

Concept of E-Learning Solutions

Increasing demand for corporate training as well as the logistical headaches associated with assembling instructors and students at a single location has been the main reason for the birth of E-learning services. In an attempt to avoid the logistical complications of on location training, many companies have opted for a virtual approach to training. Stand-alone CD-ROM modules, enterprise-wide, Web-based programs comprised of hundreds of courses custom-designed for thousands of users all fall under the concept of E-Learning. However, there're many who're skeptical of E-learning services and believes it to be just a lot of hype.

E-learning is basically a new or different form of learning. E-learning enables learners to combine traditional methods of learning with the internet facilities. In fact, some learners opt for a mix of face-to-face experiences with network-based education. However, like everything, E-Learning programs have its pros and cons. It has been found that often, E-learning provides a more individualized, self-paced, self-directed learning experience.

A main advantage of E-learning services is flexibility. With E-learning services there is no need to go to an institution in order to learn a subject or a skill. Also, you can work as well as study because E-Learning solutions don't have fixed classroom timings and therefore you've the freedom to learn at your own pace and at your convenience. E-Learning solutions also have the flexibility of being CD-ROM-based, Network-based, Intranet-based or Internet-based.

It's a hands-on learning experience without the disadvantages like crowded classrooms or a huge auditorium, where your questions and views might not get addressed. Another advantage is the amount of time you can save. With E-learning services you can save the time as well as costs that you'd otherwise spend commuting to your classes. But, you'll however miss out on the experiences that come with campus life and the social interaction that are a part of traditional classrooms and institutions. And with E-learning solutions, you can learn as and when you need it and your learning options aren't constrained or limited by your geographic location.

Another unique feature of E-learning services is that the instruction can be more personalized and flexible. Therefore it's more effective for particular kinds of learners like, shy, reflective, language challenged or slow. Within E-learning the focus is on the students and not on the instructor.

However, like everything, it has its share of disadvantages as well. For example, E-learning services that have ineffective online instructors are not very useful. Also, many instructors find it difficult to use E-learning services for traditional content. And it's at the same time more time consuming as instructors have to provide individualized feedback. Also, often there is the matter of equipment needs of students and learning providers as well as the technical training and support of learners and instructors.

Another issue that needs to be addressed is the academic honesty of online students, especially when it comes to testing and assessment. In this matter the absence of face-to-face interaction is a big set back. Another fact is that E-learning services are not available to all learners from all walks of life, because it needs special infrastructure to support it as well as new skills and responsibilities from the students. However, E-learning services provided by offshore E-Learning solutions provider companies have now become more universal, and is now giving people and organizations the competitive edge that allows them to stay ahead of the quickly varying global information economy. In the long run, this will help the Indian economy grow, since many companies have now started to outsource their E-learning services to India.

The Nature of Technology in Learning

Throughout the 20th century there were developments of the role of technology in learning. Pressey's testing machine of 1926-27 is well known but his main contribution to educational technology lay not so much in his machine as in his strong belief that an industrial revolution in education was about to dawn, bringing great benefits of more effective and more efficient learning. He pursued this dream for several decades, although he had little time for programmed learning or for teaching machines when these came along. Even his own machines were thrown away in favour of a small card with blobs of ink on it; the learner erased the blob over the answer he thought correct, and underneath was a symbol that told him whether he was right.

"We are on the threshold of an exciting and revolutionary period, in which the scientific study of man will be put to work in man's best interest. Education must play its part. It must accept that a sweeping revision of educational practices is possible and inevitable". With such evalgelising zeal did Skinner write in his 1954 article The Science of Learning and the Art of Teaching.

Skinner saw four serious shortcomings in the educational system:

- The reinforcers used were still aversive
- They were used too long after responses had been elicited
- The progression towards the required behaviour was poorly arranged
- Reinforcement was provided too infrequently.

Skinner suggested that few teachers, if any, could remedy these shortcomings working alone with a group of pupils and proposed that machines might be employed to perform most of the function the teacher could not perform, as well as some of those she could. Skinner saw programmed learning and teaching machines as part (if not all) of an overall improvement in teaching techniques. The use of technology in learning is different in its use in traditional group-based face-to-face teaching and in distance education, which is frequently individual-based and separates the learner not only from the teacher but also from the learning group.

Traditional group-based face-to-face education and training has used technology as a supplement to the teacher, and differs from distance education in which technology is a substitute for the teacher. However, in the late 1990s, with the arrival of the WWW and the provision of some universities of web based courses in place of lectures, the web has become an option on the campus as well as at a distance.

In distance education one can follow the development of a series of developments of the use of technology for teaching. The first generation uses the technology of printing and was basically the provision of print based materials for learning. A second generation added multimedia including audio, video and CD Roms to replace or supplement the print-based materials. The third generation of the 1990s was the impact of eLearning and the arrival of the WWW.

APPLICATION OF USE OF ICT IN DISTANCE EDUCATION

Application of information and communication technology especially in support of distance education activities, ads enormous value to the training. It enables and empower trainee and participating teachers to share experience, curriculum, learning materials, lesson notes, feedback and collaborative programmes. It encourages paced learning through tools such as assignment, tutorials, broadcast programmes, computers, conferences, etc.

The advantages of ICT lie in its potential for increased interaction with and between learners, speedier delivery and response times to queries and feedback on assignments, greater access to communities of teachers and quicker lead in terms for updating course materials while at the same time needing the establishment of effective quality assurance procedures). The use of ICT and CD-ROMs is becoming more common for materials developers of distance education programmes of all kinds and for supporting administrative processes and staff involved in learner– support.

Strategies for Teacher Education Programmes:

- Closing the gap between educational objectives and teacher competencies
- Teachers as learners
- Teacher education is a continuous process
- Reviewing cost-effectiveness criteria applied to teacher education
- ICT has rendered education portable, flexible, electronic, convenient and enjoyable
- School autonomy implies teacher autonomy
- An integral approach to teacher professional development.

Teaching Competencies: Teacher Education is a professional course, where a Student – Teacher wants to develop one's teaching competencies.

In distance education the STs are in service, so they have certain skills developed them and also they have a platform to practice whatever inputs are given to them in the training programme. Considering this background we have developed a 'training continuum' to develop their teaching Competencies.

Functional purpose of innovation in education is the implement of multidimensional practices in teacher education. Structural purpose is the change of basic fabric in teacher education. The visionary purpose is the continued articulation of teacher's role and the fundamental purpose is in our image of teacher as-the key reformer in education.

Any functional or structural improvement in the multidimensional practices in teacher education should be guided or judged by the image of teachers as reformer.

In institutionalized system of education direct and face--to-face teaching and instruction is great significance. The teacher and the taught share their knowledge, experiences and attitude and try to update their skills. The teacher is giver and student is taker. Since the success of educative process depends upon the efficiency of the teacher education system in the country.

The micro-teaching training program for trainees pursuing B.Ed through distance education. The trainers used the techniques of Visualization, Diagnosis, Remediation, Scaffolding and knowledge Integration. The Trainees were given exposure to online workshop, Seminars, Tele-Conferences and expert advices were sought with the help of Internet. Then their teaching competency was assessed through CCTV's. The Pros and Corns of their Teaching Method were analyzed by the Peers and by the experts. Thus the results showed that ICT helps in shaping the Teaching Competency of the teacher trainees.

CORRESPONDENCE STUDY TO DISTANCE EDUCATION

In 1982, the International Council for Correspondence Education changed its name to the International Council for Distance Education to reflect the developments in the field. With the rapid growth of new technologies and the evolution of systems for delivering information, distance education, with its ideals of providing equality of access to education, became a reality. Today there are distance education courses offered by dozens of public and private organizations and institutions to school districts, universities, the military, and large corporations. Direct satellite broadcasts are produced by more than 20 of the country's major universities to provide over 500 courses in engineering delivered live by satellite as part of the National Technological University (NTU). In the corporate sector, more than $40 billion a year are spent by IBM, Kodak, and the Fortune 500 companies in distance education programs.

What, exactly, are the prospects and promises of distance education? Desmond Keegan identified six key elements of distance education:

- Separation of teacher and learner
- Influence of an educational organization
- Use of media to link teacher and learner

- Two-way exchange of communication
- Learners as individuals rather than grouped
- Educators as an industrialized form.

Distance education has traditionally been defined as instruction through print or electronic communications media to persons engaged in planned learning in a place or time different from that of the instructor or instructors. The traditional definition of distance education is slowly being eroded as new technological developments challenge educators to reconceptualize the idea of schooling and lifelong learning.

At the same time, interest in the unlimited possibilities of individualized distance learning is growing with the development of each new communication technology.

Although educational technologists agree that it is the systematic design of instruction that should drive the development of distance learning, the rapid development of computer related technologies has captured the interest of the public and has been responsible for much of the limelight in which distance educators currently find themselves. Although the United States has seen rapid growth in the use of technology for distance education, much of the pioneering work has been done abroad.

Open Learning in the U.K.

The establishment of the British Open University in e United Kingdom in 1969 marked the beginning of the use of technology to supplement print-based instruction through well-designed courses. Learning materials were delivered on a large scale to students in three programs: undergraduates, postgraduates, and associate students. Although course materials were primarily print based, they were supported by a variety of technologies. No formal educational qualifications have been required to be admitted to the British Open University. Courses are closely monitored and have been successfully delivered to over 100,000 students. As a direct result of its success, the Open University model has been adopted by many countries in both the developed and developing world. Researchers in the United Kingdom continue to be leaders in identifying problems and proposing solutions for practitioners in the field.

The International Centre for Distance Learning, at the British Open University, maintains the most complete holdings of literature in both research and practice of international distance learning. Research studies, evaluation reports, course modules, books, journal articles, and ephemeral material concerning distance education around the world are all available through quarterly accessions lists or on line.

Distance Education in the United States

The United States was slow to enter the distance education marketplace, and when it did, a form of distance education unique to its needs evolved. Not having the economic problems of some countries or the massive illiteracy problems of developing nations, the United States nevertheless had problems of economy of delivery. Teacher shortages in areas of science, math, and foreign language combined with state mandates to rural schools produced a climate, in the late 80s, conducive to the rapid growth of commercial courses such as those offered via satellite by the TI-IN network in Texas and at Oklahoma State University. In the United States, fewer than 10 states were promoting distance education in 1987. A year later, that number had grown to two-thirds of the states, and by 1989 virtually all states were involved in distance learning programs. Perhaps the most important political document describing the state of distance education has been the report done for Congress by the Office of Technology Assessment in 1989 called Linking for Learning (Office of Technology Assessment, 1989). The report gives an overview of distance learning, the role of teachers, and reports of local, state, and federal projects. It describes the state of distance education programs throughout the United States in 1989 and highlights how technology was being used in the schools. Model state networks and telecommunication delivery systems are outlined with recommendations given for setting up local and wide-area networks to link schools. Some projects, such as the Panhandle Shared Video Network and the Iowa Educational Telecommunications Network, serve as examples of operating video networks that are both,efficient and cost effective.

COMPUTERS IN DISTANCE EDUCATION

In recent years, educators have witnessed the rapid development of computer networks, dramatic improvements in the processing power of personal computers, and striking advances in magnetic storage technology. These developments have made the computer a dynamic force in distance education, providing a new and interactive means of overcoming time and distance to reach learners.

Computer applications for distance education fall into four broad categories:

- Computer Assisted Instruction (CAI) -uses the computer as a self-contained teaching machine to present discrete lessons to achieve specific but limited educational objectives. There are several CAI modes, including: drill and practice, tutorial, simulations and games, and problem-solving.
- Computer Managed Instruction (CMI) -uses the computer's branching,

storage, and retrieval capabilities to organize instruction and track student records and progress. The instruction need not be delivered via computer, although often CAI (the instructional component) is combined with CMI.

- Computer Mediated Communication (CMC)-describes computer applications that facilitate communication. Examples include electronic mail, computer conferencing, and electronic bulletin boards.
- Computer-Based Multimedia-HyperCard, hypermedia, and a still-developing generation of powerful, sophisticated, and flexible computing tools have gained the attention of distance educators in recent years. The goal of computer-based multimedia is to integrate various voice, video, and computer technologies into a single, easily accessible delivery system.

Advantages of Computers

- Computers can facilitate self-paced learning. In the CAI mode, for example, computers individualize learning, while giving immediate reinforcement and feedback.
- Computers are a multimedia tool. With integrated graphic, print, audio, and video capabilities, computers can effectively link various technologies. Interactive video and CD-ROM technologies can be incorporated into computer-based instructional units, lessons, and learning environments.
- Computers are interactive. Microcomputer systems incorporating various software packages are extremely flexible and maximize learner control.
- Computer technology is rapidly advancing. Innovations are constantly emerging, while related costs drop. By understanding their present needs and future technical requirements, the cost-conscious educator can effectively navigate the volatile computer hardware and software market.
- Computers increase access. Local, regional, and national networks link resources and individuals, wherever they might be. In fact, many institutions now offer complete undergraduate and graduate programs relying almost exclusively on computer-based resources.

Limitations of Computers

- Computer networks are costly to develop. Although individual computers are relatively inexpensive and the computer hardware and software market is very competitive, it is still costly to develop instructional networks and purchase the system software to run them.

- The technology is changing rapidly. Computer technology evolves so quickly that the distant educator focused solely on innovation "not meeting tangible needs" will constantly change equipment in an effort to keep pace with the "latest" technical advancements.
- Widespread computer illiteracy still exists. While computers have been widely used since the 1960's, there are many who do not have access to computers or computer networks.
- Students must be highly motivated and proficient in computer operation before they can successfully function in a computer-based distance learning environment.

The Internet and Distance Education

The Internet is the largest, most powerful computer network in the world. It encompasses 1.3 million computers with Internet addresses that are used by up to 30 million people in more than fifty countries. As more and more colleges, universities, schools, companies, and private citizens connect to the Internet either through affiliations with regional not-for-profit networks or by subscribing to information services provided by for-profit companies, more possibilities are opened for distance educators to overcome time and distance to reach students.

With access to the Internet, distance educators and their students can use:

- Electronic mail (e-mail)- Like postal mail, e-mail is used to exchange messages or other information with people. Instead of being delivered by the postal service to a postal address, e-mail is delivered by Internet software through a computer network to a computer address.
- Bulletin boards- Many bulletin boards can be accessed through the Internet. Two common public bulletin boards on the Internet are USENET and LISTSERV. USENET is a collection of thousands of topically organized newsgroups, covering everything from supercomputer design to bungee cord jumping, and ranging in distribution from the whole world to single institutions. LISTSERV also provides discussion forums on a variety of topics broken out by topic or area of special interest.
- World-Wide Web (WWW)-The WWW is an exciting and innovative front-end to the Internet. Officially WWW is described as a "...wide-area hypermedia information retrieval initiative aiming to give universal access to a large universe of documents". The WWW provides Internet users with a uniform and convenient means of accessing the wide variety of resources (pictures, text, data, sound, video) available on the Internet. Popular software interfaces, such as Mosaic and Netscape, facilitate navigation and use of the WWW. The central organizing

feature of the WWW is the "home page". Every organization and even every individual user of the WWW can create a home page that contains whatever information they want to present. The hypertext capabilities of the WWW facilitate linking of information within your own home page and with all other home pages on the WWW.

Instructional Possibilities of the Internet

Distance educators can use the Internet and WWW to help students gain a basic understanding of how to navigate and take full advantage of the networked world into which they will be graduating. Some instructional possibilities of the Internet include:

- Using e-mail for informal one-to-one correspondence. Feedback from the instructor can be received more quickly than messages sent by mail. Students can read messages at their convenience and easily store them for later reference.
- Establishing a classroom bulletin board. Distant students often work in isolation without the assistance and support of fellow students. Setting up a class bulletin board can encourage student-to-student interaction. With a class computer conference, individual students can post their comments or questions to the class, and every other individual is free to respond. The conference can also be used to post all modifications to the class schedule or curriculum, assignments/tests, and answers to assignments/tests.
- Engaging students in dialogue with other students, faculty, and researchers by encouraging them to join a bulletin board(s) on topic(s) related to the class.
- Developing a classroom home page. The home page can cover information about the class including the syllabus, exercises, literature references, and the instructor's biography. The instructor can also provide links to information on the WWW that would be useful to students in the class (e.g., real research data on agricultural markets, global climate change, or space missions). Other links could access library catalogs or each student's individual home page.

Teaching Considerations

When incorporating the Internet into a distance delivered course, remember that:

- All students in a course must have Internet and WWW access to ensure equal opportunities for computer interaction and feedback. Also, convenient access to a computer at home or work may influence student success.

- Students may face the concurrent challenges of learning basic computer skills, new software, and appropriate online communication skills. Trouble-shooting student computer problems will probably become a part of normal instructional responsibilities. Setting up a specific classroom conference for ongoing discussions of specific hardware and software problems may help students to work through these problems on their own.
- Some students might hesitate to contribute to computer conferences or to send e-mail because of a lack of familiarity with the proper protocols. Encourage students to use e-mail, classroom conferences, electronic bulletin boards, and the WWW early in the course so they overcome inhibitions. Specifying a minimum number of e-mail communications per week will encourage active participation.
- Using e-mail can help the instructor provide feedback more quickly than surface mail or telephone. Prompt response generally increases student motivation and performance.
- Prompt responses might not always be appropriate. Computer conferences can foster student-to-student interaction. To ensure that this interaction is sustained, work towards a facilitative role. It might be appropriate to delay response to a query in a classroom conference in order to allow students to respond to the issue and to each other.
- Becoming familiar with the resources available on the Internet and the most effective ways to use them will be part of the instructional challenge. A number of helpful guides to the Internet and WWW are available.

PRINT IN DISTANCE EDUCATION

Print is the foundation of distance education and the basis from which all other delivery systems have evolved.

The first distance-delivered courses were offered by correspondence study, with print materials sent and returned to students by mail.

While technological developments have added to the repertoire of tools available to the distance educator, print continues to be a significant component of all distance education programs.

Advantages of Print

- Spontaneous. Print materials can be used in any setting without the need for sophisticated presentation equipment.
- Instructionally transparent. The medium of delivery should enhance, not compete with, the content for the learner's attention. If the student

reads well, the print medium is the most transparent instructional medium of all.

- Non-threatening. Reading is second nature to most students. As a result, they are easily able to focus on the content, without becoming mesmerized or frustrated by the process of reading itself.
- Easy to use. Given adequate light, print materials can be used any time and any place without the aid of supplemental resources such as electricity, viewing screen, and specially designed electronic classrooms. The portability of print is especially important for rural learners with limited access to advanced technology.
- Easily reviewed and referenced. Print materials are typically learner-controlled. As a result, the student rapidly moves through redundant sections, while focusing on areas demanding additional attention.
- Cost-effective. No instructional tool is less expensive to produce than print. In addition, facilities abound for the inexpensive duplication of these materials.
- Easily edited and revised. In comparison to technically sophisticated electronic software, print is both easy and inexpensive to edit and revise.
- Time-effective. When instructional print materials are created, the developer's primary focus remains on content concerns, not the technical requirements of the delivery system.

Limitations of Print

- Limited view of reality. Print, by its reliance on the written word, offers a vicarious view of reality. Despite the use of excellent sequential illustrations or photos, for example, it is impossible to adequately recreate motion in print.
- Passive and self-directed. Numerous studies have shown that higher learner motivation is required to successfully complete print-based courses. To a certain extent, the passive nature of print can be offset by systematic instructional design that seeks to stimulate the passive learner. Still, it takes more motivation to read a book or work through a written exercise than it does to watch a television program or participate in an audio conference with an instructor encouraging student participation and response.
- Feedback and interaction. Without feedback and interaction, instruction suffers, regardless of the delivery system in use. By nature, print materials are passive and self-directed. Even with print materials incorporating feedback mechanisms and interactive exercises, it is easy for learners to skip to the answer section.

- Dependent on reading skills. Thanks to television, most students have developed fairly good viewing skills by age four. These same children, however, often fail to develop adequate reading skills by age 12. Reading skills must often be improved. Lack of ability in this area cripples the effectiveness of even the most instructionally sound print material and must be overcome if print is to be used effectively.

Formats of Print Materials

Various print formats are available, including:

- Textbooks. As in traditionally delivered courses, textbooks are the basis and primary source of content for the majority of distance-delivered courses. While textbooks should always be critically reviewed before adoption, this is especially critical when the learner and the instructor are not in daily contact.
- Study guides. Typically, distance educators use study guides to reinforce points made during class and through the use of other delivery systems. They will often include exercises, related readings and additional resources available to the student.
- Workbooks. In a distance education context workbooks are often used to provide course content in an interactive manner. A typical format might contain an overview, the content to be covered, one or more exercises or case studies to elaborate the points being made, and a quiz or test (with answer key) for self-assessment. In addition, there is typically some form of feedback, remediation, or 'branching' loop to recycle students through the instruction as needed.
- Course syllabus. A comprehensive and well-planned course syllabus is the foundation of many distance-delivered courses. It provides course goals and objectives, performance expectations, descriptions of assignments, related readings (often by session), grading criteria, and a day-by-day overview of the material to be covered. The syllabus must be as complete as possible in order to guide the students through the course in the absence of daily contact with the instructor.
- Case studies. If written imaginatively, case studies are an extremely effective instructional tool. In fact, case studies are often designed around the limitations of print and intended to spark the students' imaginations as they place themselves in the particular case under consideration. Many case studies present a content-based scenario. They raise questions, pose alternative solutions, and then branch students to different sections of the text. There, the consequences of the selected alternative are described.

7

Computer Application in Educational Institutions

TRENDS IN COMPUTER SYSTEMS

Today's computer systems come in a variety of sizes, shapes, and computing capabilities. Rapid hardware and software developments and changing end user needs continue to drive the emergence of new models of computers, from the smallest hand-held personal digital assistant for end users, to the largest multiple-CPU mainframe for the enterprise. Categories such as mainframes, midrange computers, and microcomputers are still used to help us express the relative processing power and number of end users that can be supported by different types of computers.

In addition, experts continue to predict the merging or disappearance of several computer categories. They feel, for example, that many midrange and mainframe systems have been made obsolete by the power and versatility of client/server networks of end user microcomputers and servers.

COMPUTER GENERATIONS

It is important to realize that major changes and trends in computer systems have occurred during the major stages-or generations-of computing, and will continue into the future.

The first generation of computers developed in the early 1950s, the second generation blossomed during the late 1960s, the third generation took computing into the 1970s, and the fourth generation has been the computer technology of the 1980s and 1990s. A fifth generation of computers that accelerates the trends of the previous generations is expected to evolve as we enter the 21st century. Notice that computers continue to become smaller, faster, more reliable, less costly to purchase and maintain, and more interconnected within computer networks.

First-generation computing involved massive computers using hundreds or thousands of vacuum tubes for their processing and memory circuitry. These large computers generated enormous amounts of heat; their vacuum tubes had to be replaced frequently. Thus, they had large electrical power, air conditioning, and maintenance requirements. First-generation computers had main memories of only a few thousand characters and millisecond processing speeds. They used magnetic drums or tape for secondary storage and punched cards or paper tape as input and output media.

Second-generation computing used transistors and other solid-state, semiconductor devices that were wired to circuit boards in the computers. Transistorized circuits were much smaller and much more reliable, generated little heat, were less expensive, and required less power than vacuum tubes. Tiny magnetic cores were used for the computer's memory, or internal storage. Many second-generation computers had main memory capacities of less than 100 kilobytes and microsecond processing, speeds. Removable magnetic disk packs were introduced, and magnetic tape merged as the major input, output, and secondary storage medium for large computer installations.

Third-generation computing saw the development of computers that used integrated circuits, in which thousands of transistors and other circuit elements are etched on tiny chips of silicon. Main memory capacities increased to several megabytes and processing speeds jumped to millions of instructions per second (MIPS) as telecommunications capabilities became common. This made it possible for operating system programs to come into widespread use that automated and supervised the activities of many types of peripheral devices and processing by mainframe computers of several programs at the same time, frequently involving networks of users at remote terminals. Integrated circuit technology also made possible the development and widespread use of small computers called minicomputers in the third computer generation.

Fourth-generation computing relies on the use of LSI (large-scale integration) and VLSI (very-large-scale integration) technologies that cram hundreds of thousands or millions of transistors and other circuit elements on each chip. This enabled the development of microprocessors, in which all of the circuits of a CP are contained on a single chip with processing speeds of millions of instructions per second. Main memory capacities ranging from a few megabytes to several gigabytes can also be achieved by memory chips that replaced magnetic core memories. Microcomputers, which use microprocessor CPUs and a variety of peripheral devices and easy-to-use software packages to form small personal computer (PC), systems or client/ server networks of linked PCs and servers, are a hallmark of the fourth generation of computing, which accelerated the downsizing of computing systems. Whether we are moving into a fifth generation of computing is a subject of debated since the concept of generations may no longer fit the continual, rapid changes occurring in computer hardware, software, data, and networking technologies. But in any case, we can be sure that progress in computing will continue to accelerate, and that the development of Internet-based technologies and applications will be one of the major forces driving computing into the 21st century.

MICROCOMPUTER SYSTEMS

Microcomputers are the most important category of computer systems for end users. Though usually called a personal computer, or PC, a microcomputer is much more than a small computer for use by an individual. The computing power of microcomputers now exceeds that of the mainframes of previous computer generations at a fraction of their cost. Thus, they have become powerful networked professional work stations for end users in business. Microcomputers come in a variety of sizes and shapes for a variety of purposes. For example, PCs are available as handhled, notebook, laptop, portable, desktop, and floor-standing models. Or, based on their use, they include home, personal, professional, workstation, and multi-user systems. Most microcomputers are desktops designed to fit on an office desk, or notebooks for those who want a small, portable PC for their work activities.

Some microcomputers are powerful workstation computers (technical work-stations) that support applications with heavy mathematical computing and graphics display demands such as computer-aided design (CAD) in engineering, or investment and portfolio analysis in the securities industry. Other microcomputers are used as network servers. They are usually more powerful microcomputers that coordinate telecommunications and resource sharing in small local area networks (LANs), and Internet and intranet Web sites. Another important microcomputer category includes handheld microcomputer devices known as personal digital assistants (PDAs), designed for convenient mobile communications and computing. PDAs use touch-screens, pen-based handwriting recognition of keyboards to help mobile workers send and receive E-mail and exchange information such as appointments, to do lists, and scales contacts with their desktop PCs or Web servers.

MULTIMEDIA SYSTEMS

Multimedia PCs are designed to present you with information in a variety of media, including text and graphics displays, voice and other digitized audio, photographs, animation, and video clips. Mention multimedia, and many people think of computer video games, multimedia encyclopedias, educational videos, and multimedia home pages on the World Wide Web. However, multimedia systems are widely used in business for training employees, educating customers, making sales presentations, and adding impact to other business presentations.

The basic hardware and software requirements of a multimedia computer system depend on whether you wish to create as well as enjoy multimedia presentations.

Owners of low-cost multimedia PCs marketed for home used do not need authoring software or high-powered hardware capacities in order to enjoy multimedia games and other entertainment and educational multimedia products. These computers come equipped with a CD-ROM drive, stereo speakers, additional memory, a high-performance processor, and other multimedia processing capabilities.

People who want to create their own multimedia production may have to spend several thousand dollars to put together a high-performance multimedia authoring system. This includes a high-resolution color graphics monitor, sound and video capture boards, a high-performance microprocessor with multimedia capabilities, additional megabytes of memory, and several gigabytes of hard disk capacity.

Sound cards and video capture boards are circuit boards that contain digital signal processors (DSPs) and additional megabytes of memory for digital processing of sound and video. A digital camera, digital video camcorder, optical scanner, and software such as authoring tools and programs for image editing and graphics creation can add several thousand dollars to the star-up costs of a multimedia authoring system.

USE OF COMPUTER IN EDUCATIONAL MANAGEMENT

Technology has opened many educational doors to children, particularly to children with disabilities. Alternative solutions from the world of technology are accommodating physical, sensory, or cognitive impairments in many ways.

Much of the technology we see daily was developed initially to assist persons with disabilities. Curb cuts at streetcorners and curb slopes, originally designed to accommodate people with orthopedic disabilities, are used more frequently by families with strollers or individuals with grocery carts than by persons with wheelchairs or walkers. The optical character reader, developed to assist individuals unable to read written text, has been adapted in the workplace to scan printed documents into computer-based editable material, saving enormous amounts of data entry labor.

Technology—An Equalizer

Technology can be a great equalizer for individuals with disabilities that might prevent full participation in school, work, and the community. This is most evident in the case of individuals with mobility, hearing, or vision impairments, but is also true for individuals with limitations in cognition and perception. With technology, an individual physically unable to speak can communicate with spoken language. Using a portable voice synthesizer, a student can ask and respond to questions in the "regular" classroom, overcoming a physical obstacle that may have forced placement in a special segregated classroom or required a full-time instructional aide or interpreter to provide "a voice."

Improvements in sensor controls enable subtle motor movements to control mobility devices, such as electric wheelchairs, providing independent movement through the school and community. Text and graphics enhancement software can enlarge sections of a monitor enough to be seen by persons with vision impairments. Text can be read electronically by a digitized voice synthesizer for a person who is blind. For persons with hearing impairments, amplification devices can filter extraneous noise from the background or pick up an FM signal from a microphone on a teacher's lapel.

Word processing, editing, spellchecking, and grammatical tools commonly found in high-end software facilitate the inclusion of students with learning disabilities in regular classrooms by allowing them to keep up with much of the work. Not inconsequentially, the children often feel better about themselves as active learners.

Technology is providing more powerful and efficient tools to teachers who work with children with disabilities. These tools enable teachers to offer new and more effective means of learning while individualizing instruction to the broad range of student learning needs. Educators are using computers as tools to deliver and facilitate learning beyond drill and practice, to provide environments that accommodate learning, and to ensure enhanced and equitable learning environments to all students.

Access to the World Wide Web, email, listservs, and other electronic learning environments is common in many classrooms. In these environments, students around the world can interact in real time via onscreen messaging or video and audio transmissions. In most of these learning situations, a disability makes no difference at all.

Assistive Technology Defined

The definition of assistive technology applied to education is extremely broad, encompassing "any item, piece of equipment, or product system whether acquired commercially off the shelf, modified, or customized, that is used to increase, maintain, or improve functional capabilities of individuals with disabilities."

As a result, the potential range of AT devices is incredibly large, and both "high-tech" and "low-tech" devices are included. High-tech devices may be computers, electronic equipment, or software. Although electronically operated, high-tech devices need not be expensive, a simple low-cost switch that controls a battery-operated toy can be considered a high-tech device, as can a tape recorder. Low-tech devices are manually, not electronically, operated. This group includes devices such as pencil grips, mouth sticks, and mechanical hoists.

This definition also expands the consideration of potential educational applications with its focus on devices "used to increase, maintain, or improve the functional capabilities of persons with disabilities." As educators, we try to increase or add new academic, social, and daily living skills and knowledge to the functional capability of all children. This is a basic goal as we prepare children to take their place in society.

In the case of children with degenerative impairments, such as muscular dystrophy, educators may be working to keep children functioning at their current level. They may be striving to help students maintain their capability to function in the world. Teachers work with students to improve skills and knowledge, making existing skills and knowledge even more functional and improving fluency so that functional capabilities may be generalized into different settings.

It is critical to understand the implications of this definition to comprehend its effect on children with disabilities in our schools. It is fairly easy to understand how the definition is applied with regard to children with physical or sensory disabilities. To see a young child who had been unable to speak for her first five years say her first sentence with a speaking computer device presents an exciting and clear picture of assistive technology. The benefit of AT is also easy to comprehend when a child who cannot hear can understand his teacher's directions because real-time captioning converts the teacher's speech to text projected onto his laptop computer.

The definition of assistive technology also applies to the more difficult-to-gauge tools that teachers use to deliver and facilitate learning, including instructional applications of technology. These applications range from drill and practice tutorials to facilitated learner-based environments provided through the Internet or interactive hypermedia and multimedia-based instruction.

It is important to understand that virtually all applications of technology — tools for children to learn, as well as tools for teachers to provide learning opportunities — can be defined as assistive technology. This is true for individual children with disabilities whose disability has a primary impact on academic performance (e.g., learning disabilities) or functional performance (e.g., multiple physical and visual disabilities).

Legal and Moral Requirements

The mandate to provide assistive technology to children with special needs is grounded in the moral concerns protected by the U.S. Constitution and its amendments. *The Education for All Handicapped Children Act* (P.L. 94-142) was based on the Supreme Court's 1954 *Brown vs. Board of Education* decision that separate education was not equal education under the 14th Amendment to the Constitution. At the time the law was passed by Congress in 1975, nearly 2 million children were excluded from schools in the United States. With the legislation, the president and the Congress established a legal requirement for a "free appropriate public education in the least restrictive environment" for children with disabilities and, as a result, the field of special education began to flourish for the first time in nearly seventy-five years.

Many controversies surfaced, however, about the extent of the required educational services and the cost to society for those services. The major debates have focused on the need for a clear definition of an "appropriate" education in the least restrictive environment and the requirement to provide assistive technology devices and services to all individuals with disabilities.

'Appropriate' Education

The requirement for an "appropriate" education in the least restrictive environment has led to the development of a separate educational system designed to meet the needs of children with disabilities. Some educators contend that this is the same type of separate system that the Supreme Court found unconstitutional in 1954. These individuals suggest that all children, regardless of ability, should be educated with their neighborhood peers in their local school.

Others in favor of the special education system argue that it is necessary to meet the educational needs of all children with disabilities, particularly in the "continuum of services" mandated by the Individuals with Disabilities Act (IDEA). In their view, children must have specific intervention designed to "mainstream" them back into regular education. Without the intervention, these individuals believe that students will be doomed to continued and more significant failure. They also note that, while the goal of mainstreaming is reasonable, some children may not benefit appropriately from a full inclusion programme. Although there are many arguments on both sides of the issue, it is apparent that new technologies can provide the tools to bring more children with disabilities into "regular" educational settings. In my opinion, assistive technology will certainly mainstream more and more children in wheelchairs, children who cannot physically speak, see, or hear, and children who need computers to write, organize, think, and function educationally.

The AT Requirement

The second debate centers on the requirement to provide assistive technology to all students. The initial legislation, the Education for All Handicapped Children Act, did not require schools to provide assistive technology devices and services to individuals with disabilities. The current assistive technology mandate was created by later legislation and prompted by the technological revolution resulting from the development of the microcomputer.

Subsequent legislation passed by Congress encouraged states to develop services designed to provide assistive technology to all persons with disabilities and required provision of AT as a special education service (trained special education teachers in special classes), related service (occupational, physical, speech therapies, and other services needed to access education) or supplemental service (services necessary to maintain a child in regular education classes).

Many states have not addressed the AT issue, since assistive technology devices and services were identified as requirements only recently. This may be due to a fear of "breaking" instructional budgets by purchasing high-cost equipment in already cash-short school systems.

Concern also exists that the rapid evolution of technology creates the potential of costly investment in devices that may have a relatively short life span. A close look at the situation will show that these concerns are not well grounded, however. Schools already use extensive amounts of AT, and need only to identify it as such. Nearly any use of computers falls into this category, as do tape recorded instructions or homework, copies of notes from a classmate or teacher, switch-operated toys, drawing paper taped to table tops, as well as large pencils and crayons. All of these could be noted, as required, in Individual Education Plans (IEPs) and Individual Family Service Plans (IFSPs).

APPLICATION SOFTWARE

Application software is any tool that functions and is operated by means of a computer, with the purpose of supporting or improving the software user's work. In other words, it is the subclass of computer software that employs the capabilities of a computer directly and thoroughly to a task that the user wishes to perform.

This should be contrasted with system software (infrastructure) or middleware (computer services/ processes integrators), which is involved in integrating a computer's various capabilities, but typically does not directly apply them in the performance of tasks that benefit the user. In this context the term application refers to both the *application software* and its implementation. A simple, if imperfect analogy in the world of hardware would be the relationship of an electric light bulb (an application) to an electric power generation plant (a system). The power plant merely generates electricity, not itself of any real use until harnessed to an application like the electric light that performs a service that benefits the user.

Typical examples of 'software applications' are word processors, spreadsheets, media players and database applications. Multiple applications bundled together as a package are sometimes referred to as an application suite. Microsoft Office, OpenOffice.org, and iWork, which bundle together a word processor, a spreadsheet, and several other discrete applications, are typical examples. The separate applications in a suite usually have a user interface that has some commonality making it easier for the user to learn and use each application.

And often they may have some capability to interact with each other in ways beneficial to the user. For example, a spreadsheet may be embedded in a word processor document even though it has been created in a separate spreadsheet application.

User-written software tailors systems to meet the user's specific needs. User-written software include spreadsheet templates, word processor macros, scientific simulations, graphics and animation scripts. Even email filters are a kind of user software. Users create this software themselves and often overlook how important it is. In some types of embedded systems, the application software and the operating system software may be indistinguishable to the user, as in the case of software used to control a VCR, DVD player or microwave oven. It is important to note that this definition may exclude some applications that may exist on some computers in large organizations.

Terminology

The exact delineation between system software such as operating systems and application software is not precise, however, and is occasionally subject to controversy. For example, one of the key questions in the United States v. Microsoft antitrust trial was whether Microsoft's Internet Explorer web browser was part of its Windows operating system or a separable piece of application software. As another example, the GNU/Linux naming controversy is, in part, due to disagreement about the relationship between the Linux kernel and the operating systems built over this kernel. In computer science, an application is a computer programme designed to help people perform a certain type of work. An application thus differs from an operating system (which runs a computer), a utility (which performs maintenance or general-purpose chores), and a programming language (with which computer programmes are created).

Depending on the work for which it was designed, an application can manipulate text, numbers, graphics, or a combination of these elements. Some application packages offer considerable computing power by focusing on a single task, such as word processing; others, called integrated software, offer somewhat less power but include several applications.

COMPUTER FOR INSTRUCTION, COMPUTER AIDED INSTRUCTION (CAI)

Main

The use of computers in education started in the 1960s. With the advent of convenient microcomputers in the 1970s, computer use in schools has become widespread from primary education through the university level and even in some preschool programs.

Instructional computers are basically used in one of two ways: either they provide a straightforward presentation of data or they fill a tutorial role in which the student is tested on comprehension. If the computer has a tutorial program, the student is asked a question by the computer; the student types in an answer and then gets an immediate response to the answer. If the answer is correct, the student is routed to more challenging problems; if the answer is incorrect, various computer messages will indicate the flaw in procedure, and the program will bypass more complicated questions until the student shows mastery in that area.

There are many advantages to using computers in educational instruction. They provide one-to-one interaction with a student, as well as an instantaneous response to the answers elicited, and allow students to proceed at their own pace. Computers are particularly useful in subjects that require drill, freeing teacher time from some classroom tasks so that a teacher can devote more time to individual students. A computer program can be used diagnostically, and, once a student's problem has been identified, it can then focus on the problem area. Finally, because of the privacy and individual attention afforded by a computer, some students are relieved of the embarrassment of giving an incorrect answer publicly or of going more slowly through lessons than other classmates.

There are drawbacks to the implementation of computers in instruction, however. They are generally costly systems to purchase, maintain, and update.

There are also fears, whether justified or not, that the use of computers in education decreases the amount of human interaction.

One of the more difficult aspects of instructional computers is the availability and development of software, or computer programs. Courseware can be bought as a fully developed package from a Software company, but the program provided this way may not suit the particular needs of the individual class or curriculum.

A courseware template may be purchased, which provides a general format for tests and drill instruction, with the individual particulars to be inserted by the individual school system or teacher.

The disadvantage to this system is that instruction tends to be boring and repetitive, with tests and questions following the same pattern for every course. Software can be developed in-house, that is, a school, course, or teacher could provide the courseware exactly tailored to its own needs, but this is expensive, time-consuming, and may require more programming expertise than is available.

Computer-Aided Instruction

Computer-Aided Instruction (CAI), diverse and rapidly expanding spectrum of computer technologies that assist the teaching and learning process. CAI is also known as computer-assisted instruction. Examples of CAI applications include guided drill and practice exercises, computer visualization of complex objects, and computer-facilitated communication between students and teachers.

The estimated number of computers in American schools has risen from 1 for every 125 students in 1981 to 1 for every 4 students in elementary and secondary school in 2006. In some schools, pupils are given their own laptop personal computers for homework and instruction. Western European, Canadian, Australian, South Korean, and Japanese schools are likewise highly computerized. Access to computers has also grown in schools in India and China. International groups have joined in efforts to provide computers to students in developing countries in Africa and elsewhere.

Types of CAI

Information that helps teach or encourages interaction can be presented on computers in the form of text or in multimedia formats, which include photographs, videos, animation, speech, and music. The guided drill is a computer program that poses questions to students, returns feedback, and selects additional questions based on the students' responses. Recent guided drill systems incorporate the principles of education in addition to subject matter knowledge into the computer program.

Computers also can help students visualize objects that are difficult or impossible to view. For example, computers can be used to display human anatomy, molecular structures, or complex geometrical objects.

Exploration and manipulation of simulated environments can be accomplished with CAI—ranging from virtual laboratory experiments that may be too difficult, expensive, or dangerous to perform in a school environment to complex virtual worlds like those used in airplane flight simulators.

CAI tools, such as word processors, spreadsheets, and databases, collect, organize, analyze, and transmit information. They also facilitate communication among students, between students and instructors, and beyond the classroom to distant students, instructors, and experts.

CAI systems can be categorized based on who controls the progression of the lesson. Early systems were linear presentations of information and guided drill, and control was directed by the author of the software. In modern systems, and especially with visualization systems and simulated environments, control often rests with the student or with the instructor.

This permits information to be reviewed or examined out of sequence. Related material also may be explored. In some group instructional activities, the lesson can progress according to the dynamics of the group.

Advantages and Disadvantages

CAI can dramatically increase a student's access to information. The program can adapt to the abilities and preferences of the individual student and increase the amount of personalized instruction a student receives. Many students benefit from the immediate responsiveness of computer interactions and appreciate the self-paced and private learning environment. Moreover, computer-learning experiences often engage the interest of students, motivating them to learn and increasing independence and personal responsibility for education. Although it is difficult to assess the effectiveness of any educational system, numerous studies have reported that CAI is successful in raising examination scores, improving student attitudes, and lowering the amount of time required to master certain material. While study results vary greatly, there is substantial evidence that CAI can enhance learning at all educational levels.

In some applications, especially those involving abstract reasoning and problem-solving processes, CAI has not been very effective. Critics claim that poorly designed CAI systems can dehumanize or regiment the educational experience and thereby diminish student interest and motivation.

Other disadvantages of CAI stem from the difficulty and expense of implementing and maintaining the necessary computer systems. Some student failures can be traced to inadequate teacher training in CAI systems. Student training in the computer technology may be required as well, and this process can distract from the core educational process. Although much effort has been directed at developing CAI systems that are easy to use and incorporate expert knowledge of teaching and learning, such systems are still far from achieving their full potential.

History of Computer-Aided Instruction

In the mid-1950s and early 1960s a collaboration between educators at Stanford University in California and International Business Machines Corporation (IBM) introduced CAI into select elementary schools. Initially, CAI programs were a linear presentation of information with drill and practice sessions.

These early CAI systems were limited by the expense and the difficulty of obtaining, maintaining, and using the computers that were available at that time. Programmed Logic for Automatic Teaching Operations (PLATO) system, another early CAI system initiated at the University of Illinois in the early 1960s and developed by Control Data Corporation, was used for higher learning. It consisted of a mainframe computer that supported up to 1000 terminals for use by individual students.

By 1985 over 100 PLATO systems were operating in the United States. From 1978 to 1985 users logged 40 million hours on PLATO systems. PLATO also introduced a communication system between students that was a forerunner of modern electronic mail (messages electronically passed from computer to computer). The Time-shared Interactive Computer-Controlled Information Television (TICCIT) system was a CAI project developed by Mitre Corporation and Brigham Young University in Utah. Based on personal computer and television technology, TICCIT was used in the early 1970s to teach freshman-level mathematics and English courses.

With the advent of cheaper and more powerful personal computers in the 1980s, use of CAI increased dramatically. In 1980 only 5 percent of elementary schools and 20 percent of secondary schools in the United States had computers for assisting instruction. Three years later, both numbers had roughly quadrupled, and by the end of the decade nearly all schools in the United States, and in most industrialized countries, were equipped with teaching computers. A crucial development for CAI has been the vast expansion of the Internet and the World Wide Web, consortiums of interlinked computers. By connecting millions of computers worldwide, these networks enable students to access huge stores of information, which greatly enhances their research capabilities.

APPLICATION SOFTWARE CLASSIFICATION

There are many subtypes of application software:

- *Enterprise software* addresses the needs of organization processes and data flow, often in a large distributed environment. (Examples include Financial, Customer Relationship Management, and Supply Chain Management). Note that Departmental Software is a sub-type of Enterprise Software with a focus on smaller organizations or groups within a large organization. (Examples include Travel Expense Management, and IT Helpdesk)
- *Enterprise infrastructure software* provides common capabilities needed to support Enterprise Software systems. (Examples include Databases, Email servers, and Network and Security Management)
- *Information worker software* addresses the needs of individuals to create and manage information, often for individual projects within a department, in contrast to enterprise management. Examples include time management, resource management, documentation tools, analytical, and collaborative. Word processors, spreadsheets, email and blog clients, personal information system, and individual media editors may aid in multiple information worker tasks.
- *Content access software* is software used primarily to access content without editing, but may include software that allows for content editing. Such software addresses the needs of individuals and groups to consume digital entertainment and published digital content. (Examples include Media Players, Web Browsers, Help browsers, and Games)
- *Educational software* is related to content access software, but has distinct requirements for delivering evaluations (tests) and tracking progress through material. It is also related to collaboration software in that many Educational Software systems include collaborative capabilities.
- *Simulation software* are computer software for simulation of physical or abstract systems for either research, training or entertainment purposes.
- *Media development software* addresses the needs of individuals who generate print and electronic media for others to consume, most often in a commercial or educational setting. This includes Graphic Art software, Desktop Publishing software, Multimedia Development software, HTML editors, Digital Animation editors, Digital Audio and Video composition, and many others.
- *Product engineering software* is used in developing hardware and software products. This includes computer aided design (CAD), computer aided

engineering (CAE), computer language editing and compiling tools, Integrated Development Environments, and Application Programmer Interfaces.

Computer Applications

Doctoral programmes are centered around computer graphics and experimental approaches to the design of fractals. Applications have been developed for the design of patterns in carpet and textile industry. Applications have also been developed in discrete dynamical systems. Currently the department is investigating Fractal Theory in improving performance of Fractal Antennas and for noise mitigation in communication systems.

The Department also works in Computer Networking and Wireless Communication with the objective to improve the Quality of Service (QoS) by reducing congestion in high-speed communication. Currently work is in progress in developing new techniques for fast communication after a disaster has occurred.The department has produced five research papers in the last two years: three in international journals and two in national journals. To keep itself abreast of latest developments in computer applications, the Department makes extensive efforts to develop partnerships with industry. We have partnered with IBM to offer an intensive training programme for teachers on 'IT Infrastructure Management' and DB2. We also have collaboration with Infosys under their Infosys Connect Programme to offer intensive training for our students.

The Teaching Learning Process at the MCA level uses the latest electronic teaching aids and is characterized by a tight coupling between lectures, tutorial and laboratory work. The former contributes to interactive classroom learning where the student is able to discuss with the teacher and peer students. The latter enables planned learning and optimization of student teacher time to maximize learning.

Types of Educational Institutions

Computer-Integrated Surgical Systems and Technology Engineering Research Center: Developing novel computing methods, interfacial technologies, and computer-integrated surgical systems to revolutionize surgical procedures in the 21st century.

The impact of Computer-Integrated Surgical Systems and Technology (cisst) on medical care procedures within the next 20 years will be as significant as the impact of Computer-Integrated Manufacturing (CIM) Systems and Technology has been on manufacturing over the past 20 years. A novel partnership between human surgeons and machines, made possible by advances in computing and engineering technology, will overcome many of the limitations of traditional surgery. By extending human surgeons' ability to plan and carry out surgical interventions more accurately and less invasively, Computer Integrated Surgery (CIS) systems will address a vital national need to greatly reduce costs, improve clinical outcomes, and improve the efficiency of health care delivery. Further, the combination of consistent execution, patient and task models, and logging of intraoperative and outcome data made possible by CIS systems will produce the same sort of process learning advantages for surgery that have been realized in semiconductor CIM systems.

IMPACT OF COMPUTER SCIENCES ON MODERN COMMUNITIES

Development and progress in the field of Computer Sciences has influenced the life and livings of the modern Communities and Society in so many ways like below:

Construction of Buildings and Residential Colonies

On Account of the availability of a variety of load bearable and non-load bearable material through the development in Computer Sciences, the modern communities look modern in terms of the construction of their buildings in the form of business centers, offices and residential colonies.

Transportation and Communication Systems

Development in Computer Sciences has been able to provide the latest available transportation and communi-cation systems to the modern communities. The distances are no more a barrier for the people living at the farthest distances of the globe.

Modernization of the Systems

The development in the Computer Sciences have modernized the sources of availability of food stuff in the shape of modernization of the method of farming, poultry farming, cattle beading, fisheries, bee keeping etc. It has resulted in multiplying the production of the food stuff as well as reducing the complexities or manual labour. The food stuff cannot be better preserved through the modern techniques available as a result of invention and discoveries in Computer Sciences. Development in Computer Sciences is helping the modern communities to take care of their water resources. It has provided artificial irrigation means as well as availability of drinking water with the construction of big water reservoirs, dams and sophisticated distribution system. It has provided big plants and simple household gadgets for the availability of pure drinking water to the modern communities. It has also provided means to have artificial rains and cultivation of water for providing additional sources of water to the modern communities.

Modern Means for the Entertainment and Leisure Time Hobbies

Development in Computer Science has provided modern and methods for the entertainment and uplifting of leisure to the modern communities. Radio, Television Video, Films and Computer services have taken a total command of providing entertainment and leisure time hobbies to the modern communities.

Health Care and Treatment of Diseases

Development in Computer Sciences have helped much in taking care of the health including treatment of illness and diseases of the members of the modern communities. It has provided better knowledge and information for the prevention and care of the diseases as well maintenance of good physical and mental health through its wider network of information technology. The dreaded diseases are now no more so dreaded as happened to be in the past. With the vast discoveries and invention in the field of health and medical sciences as well as tremendous progress in chemical sciences. Modern communities can avail the latest treatment of the diseases and look after of their health.

Development of Inter-relationship and Dependence

Development of Computer Sciences are responsible for making the modern communities too much inter-related and inter-dependent. It has given birth to the phenomenon of globalization in every aspect—physical, mental, emotional social, cultural and ethical of the behaviour of teach and every person belonging to the modern communities of this globe.

What happens to a community of the world, residing at any farthest corner of the world equally affects the working and behaviour of the other segment of the world communities. We will discuss in detail such globalization impact of Computer Sciences soon in this very chapter.

With all what has been said above, we should not conclude the developments in Computer Sciences are always bound to cast positive and desirable impact well being and progress of the modern communities. If handled improperly and utilized destructively these can yield bitter and horrifying results. Such negative impact of the development Computer Sciences on modern communities may be summarized as below:

* Too much urbanization of the communities.
* Causing heavy pollution of every sort like air pollution, water pollution, noise pollution, cultural pollution etc.
* Inequitable distribution of wealth and other material comforts in the population.

* Abolishment of the concept and existence of the harming of health and welfare of the people.
* Development of the weapons of destruction and their unmindful application.
* Side effects of the fertilizers, chemicals, pesticides insecticides used in growing foodstuff and killing harmful bio-stuff.
* Neglect of moral values and social responsibilities at the cost of material development and individualism.

ROLE OF COMPUTER SCIENCE IN THE PROCESS OF GLOBALIZATION

The global society or globalization is something, which is an accepted reality of the modern age. The development of science and technology especially related to the field of Computer Sciences is responsible to a great extent for accelerating the process of globalization.

A few of such inventions and discoveries related to the subject of Computer Science may be named as below:

* Development of printing material and communi-cation through written verbal material.
* Development of transport system with the discoveries and inventions related to travel on land surface, air, sea and space.
* Development of communication network with the inventions and discoveries like telephone, wireless, telegraph, teleprompter, radio, television, video, camera, photography still and movie pictures, satellite communication, mobile telephony, tele-conferencing, video conferencing etc.
* Development of modern sophisticated communi-cation system with the help of computer technology, internet access facilitating, e-mail and audio-visual transmission through web camera and web sites coupled with audio listening and speaking devices.
* The great invention and discoveries mentioned as above in the field of transport and communication services brought out by Computer Sciences have helped in the process of globalization in the ways narrated as below :

(i) Travelling from one corner of the globe to the other has become too convenient and less time consuming. It has provided needed mobility to the individuals or group as a whole for the interaction with the rest of the world.

(ii) It is quite easy and speedy as well as to transport commodities from one corner of the globe to the other. That is why, it is no wonder to get benefit of the productions of the commodities grown or manufactured in one part of the globe to the other parts in no time with a reduced cost and labour.

(iii) We can be in touch with the individuals, organi-zations and events belonging to the farthest corners of the globe with the help of the much development means and tools of communication. The distance has to barriers at all for any type of communication—audio as well as visual.

The facilities available as above in terms of removing barriers of distances have invariably brought together the individuals, groups and nations of the world in the form of a single knitted unit truly in the spirit of "Basudheve Kutumbkam" i.e. "the whole world is my family".

As a result the people of the world have a close interaction, inter-dependence and inter-relationship with each other irrespective of their living and working distances.

What happens with an individual, community, region or a country now affects in one way or the other the rest of the world. Poverty of a region is now the concern of the other events and political turmoil.

As a result you may find that terrorism has become global, trafficking in drugs or human beings has become global, sexually transmitted and also other types of viral and infectious diseases have become global.

One the other hand we now have a tremendous increase in the speed or globalization at the economic as well as socio-cultural exchange fronts. It has resulted in the process of denationalization of market, politics and legal systems for paving the way of the so-called global economy.

The emergence of the phenomenon of global trade and commerce has now provided enormous opportunities for an organization or company belonging to a particular part of the global to establish itself in the foreign market.

It is easy for her now to adopt first her product or services to the final users linguistic and cultural requirements and then take advantages of the internet revolution for establishing a virtual presence on the international market place with a multilingual corporate website or even resorting to an electronic business, trade and commerce according to the mutual convenience of the business/trade partners.

In this way globalization has resulted in intertwining the fates of the individuals, community and nations of the world and this dream has become true only on account of the services provided by the numerous inventions and discoveries made in the field of Computer Sciences.

COMPUTER ASSISTED INSTRUCTION (CAI)

Individually paced instruction and frame-based, computer-aided instruction comprised early attempts to provide adaptive instruction and, although successful for some types of learning, fell short because their learning environments had low fidelity and their ability to adpt was limited to branching between static screen (Murray 1998).

In the sixties the first attempts to use computers in educations were based on rather behavioristic theories with emphasis on feedback and reinforcement actions (Gazzaniga & Scarafiotti 1997).

- The teaching path was fixed and linear.
- The communication style was monodirectional (from the computer to the student) and imperative.
- Individuality was restricted to the amount of time spent in the learning process.
- The CAI programs proved useful above all for training.
- Most severe criticism: the rigidness based on the action/reaction principle.

CAI refers to computer programs that provide drill and practice exercises while CMI refers to programs that evaluate and diagnose students' needs, guide them though the next step in their learning, and record their progress. Both CAI and CMI can be used with little teacher intervention. CEI, on the other hand, requires the teacher to be involved in planning and helping to carry out learning activities.

COMPUTER MANAGED LEARNING (CML)

What is Computer Managed Learning?

Computer Managed Learning (CML) is a learning system that allows you to take university-level courses outside of a classroom environment, but within a structured timetable.

Study the assigned material on your own and complete a series of computerized unit tests, (a unit test typically covers one to three chapters), plus midterm and final exams. Some courses may also require completion of an assignment or term paper.

- Full-time students use CML to manage timetable conflicts or to get ahead in their studies.
- Part-time students find CML a convenient way to fit a university-level course into their busy schedules.
- Successful CML students are independent learners, have good time management skills and are self-motivated.

TELE/VIDEO CONFERENCING

It is when people who can't meet in the one location will use cameras and TVs to see each other and have their meeting, although they could be anywhere.

A videoconference is a set of interactive telecommunication technologies which allow two or more locations to interact via two-way video and audio transmissions simultaneously.

It has also been called visual collaboration and is a type of groupware. It differs from videophone in that it is designed to serve a conference rather than individuals.

The core technology used in a videoteleconference (VTC) system is digital compression of audio and video streams in real time. The hardware or software that performs compression is called a codec (coder/decoder). Compression rates of up to 1:500 can be achieved.

The resulting digital stream of 1s and 0s is subdivided into labelled packets, which are then transmitted through a digital network of some kind (usually ISDN or IP).

The use of audio modems in the transmission line allow for the use of POTS, or the Plain Old Telephone System, in some low-speed applications, such as videotelephony, because they convert the digital pulses to/from analog waves in the audio spectrum range.

The other components required for a VTC system include:

- Video input: video camera or webcam
- Video output: computer monitor, television or projector
- Audio input: microphones
- Audio output: usually loudspeakers associated with the display device or telephone
- Data transfer: analog or digital telephone network, LAN or Internet.

There are basically two kinds of VTC systems:

1. Dedicated systems have all required components packaged into a single piece of equipment, usually a console with a high quality remote controlled video camera. These cameras can be controlled at a distance to pan left and right, tilt up and down, and zoom. They became known as PTZ cameras. The console contains all electrical interfaces, the control computer, and the software or hardware-based codec. Omnidirectional microphones are connected to the console, as well as a TV monitor with loudspeakers and/or a video projector. There are several types of dedicated VTC devices:
    a. Large group VTC are non-portable, large, more expensive devices used for large rooms and auditoriums.

b. Small group VTC are non-portable or portable, smaller, less expensive devices used for small meeting rooms.
c. Individual VTC are usually portable devices, meant for single users, have fixed cameras, microphones and loudspeakers integrated into the console.

2. Desktop systems are add-ons (hardware boards, usually) to normal PCs, transforming them into VTC devices. A range of different cameras and microphones can be used with the board, which contains the necessary codec and transmission interfaces. Most of the desktops systems work with the H.323 standard. Videoconferences carried out via dispersed PCs are also known as e-meetings.

8

Modern Education

It is useful, at the outset, to differentiate between "integrated education", as a separate concept or process, from "holistic education." Providing a balanced combination of the main domains in a curriculum that would serve proper development of physical, intellectual and spiritual aspects or dimensions of the learner's capabilities is usually referred to as holistic education.

The purpose is to facilitate balanced growth and development of the totality of the individual's potentialities. While physical and intellectual developments are usually the focus of the educational curricula, spiritual and moral development is either receiving a rather cursory treatment or left to be taken care by the family or other institutions in the community. Since holistic education is attending to all aspects of personality it is sometimes called "global education."

Integrated education is usually bringing together, in one combined process, more than one element, subject, method or stream in the educational process. If we mix teaching history and geography, chemistry and biology, formal and informal education, or theoretical and practical education we are implementing integrated education. In the same manner holistic education may also be delivered as an integrated education. For example moral education can be delivered by integrating it into various parts of the curriculum such as literature, history, social studies, civics, sport, home economics and life sciences.

Each one of these two approaches has its own domain of application and its own benefits in the delivery system of education. In this presentation, without disregarding the value and function of the holistic approach in education, we want to consider a wider and multi-dimensional concept of integrated education. In order to respect human rights equally throughout the world, regardless of nationality, ethnicity, creed, and geographical location, reforms in education aimed at improving and advancing the quality in education should, as much as possible, be devised in a manner that it could be implemented globally. That is my view of global integrated education.

Therefore, it is different from global integrated education, as another designation for holistic education.

Challenging Issues

The ever-increasing number of subjects, and variety of activities, that are introduced into school curricula has created a theoretical program that sounds very interesting on paper. However, it has proven practically impossible to be fully implemented with desirable quality. The old type curricula included fewer and more basic subjects.

The learners were expected to master those subjects mostly by memorizing materials and mastering a few skills. Teacher training and teaching were rather simple and straightforward. Teachers could do their jobs single-handedly. In other words they did not need the assistance and collaboration of other staff members in order to do their job.

This is no more the case. The content of the curriculum has increased and includes such a variety of items and activities that teaching is becoming practically a kind of teamwork rather than a single person activity. Various expertise and skills have to be employed in order to perform what is expected and prescribed in modern education.

Various developments have caused this change. On one hand both parents being engaged in working outside the home leaves very little time for them to attend to the teaching and educating of their children. Grandparents are rarely living with their grandchildren. It is more and more expected that what was taught at home be covered by the school curriculum. Modern technology, environmental concerns, new health hazards, have all added new subjects to the curriculum. Teacher training programs are not providing knowledge and skills needed in all the subjects that are being added to the school curricula. At the same time the total time that teachers and students are at the school, and the time spent on homework at home, compared with what used to be the standard in the past, is reduced. Television programs, computer games and other entertaining activities absorbs considerably more of the after school time of students. Innovations and rapid changes of knowledge and technology, as well as the development of new gadgets must be also added to the above factors.

Curriculum development and implementation has become a very complex and hard-to-manage activity. There is need to employ more talented and capable staff to manage such rapid changes at the three levels of curriculum development, preparation of needed curricular materials including educational technology, and educational personnel. However it is becoming harder than ever to compete with other job opportunities that are more attractive and are rapidly absorbing the talents needed in the field of education.

The demand by more people at various age levels for more education on one hand, and the population explosion that results in a comparatively smaller contingent of older generations having to serve a much larger contingent of younger generations on the other hand, have caused an acute shortage of manpower for delivery system in education. Therefore, larger number of learners has to be taught by a smaller number of instructors.

ICT IN EDUCATION

Globalization and technological change—processes that have accelerated in tandem over the past fifteen years—have created a new global economy "powered by technology, fueled by information and driven by knowledge." The emergence of this new global economy has serious implications for the nature and purpose of educational institutions. As the half-life of information continues to shrink and access to information continues to grow exponentially, schools cannot remain mere venues for the transmission of a prescribed set of information from teacher to student over a fixed period of time. Rather, schools must promote "learning to learn,": i.e., the acquisition of knowledge and skills that make possible continuous learning over the lifetime. "The illiterate of the 21st century," according to futurist Alvin Toffler," will not be those who cannot read and write, but those who cannot learn, unlearn, and relearn."

Concerns over educational relevance and quality coexist with the imperative of expanding educational opportunities to those made most vulnerable by globalization—developing countries in general; low-income groups, girls and women, and low-skilled workers in particular. Global changes also put pressure on all groups to constantly acquire and apply new skills. The International Labour Organization defines the requirements for education and training in the new global economy simply as "Basic Education for All", "Core Work Skills for All" and "Lifelong Learning for All". 3 Information and communication technologies (ICTs)—which include radio and television, as well as newer digital technologies such as computers and the Internet—have been touted as potentially powerful enabling tools for educational change and reform. When used appropriately, different ICTs are said to help expand access to education, strengthen the relevance of education to the increasingly digital workplace, and raise educational quality by, among others, helping make teaching and learning into an engaging, active process connected to real life.

However, the experience of introducing different ICTs in the classroom and other educational settings all over the world over the past several decades suggests that the full realization of the potential educational benefits of ICTs is not automatic.

The effective integration of ICTs into the educational system is a complex, multifaceted process that involves not just technology—indeed, given enough initial capital, getting the technology is the easiest part!—but also curriculum and pedagogy, institutional readiness, teacher competencies, and long-term financing, among others. This primer is intended to help policymakers in developing countries define a framework for the appropriate and effective use of ICTs in their educational systems by first providing a brief overview of the potential benefits of ICT use in education and the ways by which different ICTs have been used in education thus far.

Second, it addresses the four broad issues in the use of ICTs in education—effectiveness, cost, equity, and sustainability. The primer concludes with a discussion of five key challenges that policymakers in developing countries must reckon with when making decisions about the integration of ICTs in education, namely, educational policy and planning, infrastructure, capacity building, language and content, and financing.

What are ICTs and what types of ICTs are commonly used in education? ICTs stand for *information and communication technologies* and are defined, for the purposes of this primer, as a "diverse set of technological tools and resources used to communicate, and to create, disseminate, store, and manage information." These technologies include *computers, the Internet, broadcasting* technologies (radio and television), and telephony.

In recent years there has been a groundswell of interest in how computers and the Internet can best be harnessed to improve the efficiency and effectiveness of education at all levels and in both formal and non-formal settings. But ICTs are more than just these technologies; older technologies such as the telephone, radio and television, although now given less attention, have a longer and richer history as instructional tools. For instance, radio and television have for over forty years been used for open and distance learning, although print remains the cheapest, most accessible and therefore most dominant delivery mechanism in both developed and developing countries. The use of computers and the Internet is still in its infancy in developing countries, if these are used at all, due to limited infrastructure and the attendant high costs of access.

Moreover, different technologies are typically used in combination rather than as the sole delivery mechanism. For instance, the Kothmale Community Radio Internet uses both radio broadcasts and computer and Internet technologies to facilitate the sharing of information and provide educational opportunities in a rural community in Sri Lanka. The Open University of the United Kingdom (UKOU), established in 1969 as the first educational institution in the world wholly dedicated to open and distance learning, still relies heavily on print-based materials supplemented by radio, television and, in recent years, online programming. Similarly, the Indira Gandhi National Open University in India combines the use of print, recorded audio and video, broadcast radio and television, and audioconferencing technologies.

MULTI-DIMENSIONAL APPROACH TO INTEGRATED EDUCATION

In order to cope with this highly complex situation we need to examine the possibilities for expanding the concept and the scope of integrated education and try to make it multi-dimensional. Let us look at some examples.

Integration of School Subjects

Combining and integrating school subjects will reduce the number of subjects to be included in the curriculum. For example one subject to be called "Our Environment" could be a combination of physical, political and economic geography together with environmental protection and some aspects of civics.

Another subject to be called "Our Health" could be a combination of biology, hygiene, nutrition, and some aspect of moral education.

Integration of at-school Education with Self-learning at Home

Modern educational technology is gradually making it possible to learn some subjects through various "distance education" or self-learning methods. This approach if properly devised and coordinated will reduce the burden of classroom teaching.

Integrating School- education with Out-of-school Education

There are various opportunities at home and in the community that can provide certain knowledge and skills that are normally offered by the school. Public libraries, summer jobs, assuming some responsibilities at home, services offered by public agencies and institutions can be enlisted to cover a portion of what is now expected to be included in school curriculum. This will be a kind of integration of formal and non-formal education.

Integrating Education and Learning with Recreational and Entertainment Activities

Computer games, television programs, educational video and DVD, some sport activities, summer camps, boy and girl scouts, are but a few examples of how education can be integrated with recreation and entertainment.

Integrating Academic and Vocational Education

Dividing the education process into two separate streams by separating ordinary secondary level education from vocational training, has resulted in bringing up two groups of individuals, thus dividing the society into tow camps: in one camp are those who utilize their hands namely their physical abilities and in the other camp are those who are expected to use their heads namely intellectual capabilities. The former group is not expected to use their intellectual potential and the latter are deprived of the many advantages of manual dexterity and practical skills. To give the benefits of both types of education to each individual and bringing up more balanced and capable members of society demands integrating the two streams of education into one which provides every person with the opportunity to actualize all his or her potential aptitudes and capabilities.

In short, implementation of such changes in the delivery system of education contributes to creating a learning and educational environment at home and in the community that will help the students to be surrounded, at all times, by opportunities that will be encouraging, supplementing and strengthening their educational development.

MODERN SYSTEMS OF EDUCATION

But let us first realise two fundamental differences between Ancient and Modern Systems of Education in their relation to the State, one of them prevailing alike in India and in Britain, and the other peculiar to India. In the Ancient System of India, Education and Culture were self-controlled, and while the State, the organised Nation, profited by them and from them drew its dignity, its religion, its morality, its effectiveness, and its consequent efficiency, the Legislative and Executive Departments of its Government exercised over them no control, and did not interfere with their management.

Kings built Universities and bestowed on them wealth, but claimed in them no authority. A Monarch might enter into the Convocation of a University, but no one rose to greet him and he took his seat like any other visitor; but on the entrance of its Head, the "Venerable of Venerables," all rose and turned their faces towards him and in silence awaited his words. The University was the Temple of Learning, and the learned were its only Hierophants.

When Learning visited Royalty, when a Wise One entered a Court, even Shri Krishna descended from His throne and bowed at the feet of the Sage. In the Modern System, Education is under the control of a Government Department, the Legislature makes laws for it, the Executive appoints its Directors, or the Ministers, who are really its masters, sends its Inspectors into its Schools and Colleges, and puts the Educators into a steel-frame, which it misnames efficiency, This is now alike in East and West. But in India, where Kings had been its nursing fathers and had poured out their treasures at its feet, the foreign Government ignored the Ancient System, and, as its Rule spread, Education and Culture died of starvation in the kingdoms which became provinces.

The splendid incheritance from the Indian Past—Hindu, Buddhist and Muslim—disappeared, leaving only the Schools of Pandits, maintained by Indian Princes, or by the reverent charity of the Hindus, till but one University, that of Nadiya, survived; the Temple and Masjid schools remained for a while and the muffasal village school—that which the East India Company, on being compelled by the British Parliament to spend a lakh on Education, called in 1814, "this venerable and benevolent institution of the Hindus," after the testimony of Sir Thomas Munro in 1813, that there were "schools established in every village". The E.I.Co. ascribed to these "the general intelligence of the natives as scribes and accountants". Dr. John Matthai, in his *Village Administration in British India,* says that "when the British took possession of the country," they found in most parts of the country (except western and central India) that "there existed a widespread system of national Education". Even in 1838, Adam's Reports show a similar state of things in Bengal. He reports the results of an enquiry, held in 1835-1838, made in typical districts of the Presidency, and found both Toles and Madrasahs (High Schools) and Pathashalas and Maktabs (schools attached to Temples and Masjids). The colleges were found, he writes, in "all the large villages as in the towns. The age of the scholars was from about five or six to sixteen. The curriculum included reading, writing, the composition of letters, and elementary arithmetic and accounts, either commercial, or agricultural, or both". I may add that in the village schools "Elementary Arithmetic" included multiplication tables not of only 12 × 12, but up to 20 × 20. The schools however continued to diminish in number. The *Quinquennial Review* for 1907-1912, shows 2,051 Madrasahs in 1907 against 1,446 in 1912, and 10,504 Masjid Schools in 1907 against 8,288 in 1912. Let me pause for a moment on the age of the scholars mentioned above.

In the old days, the education of the child up to the age of seven seems to have been more in the home than in the school. From seven to sixteen, the boy was to be taught and trained in school, and then to pass on to the University. The stage of infancy ends at seven, and up to that age, the body should be the first care, and lessons should be in the form of play, and great freedom of choice should be given to the little ones. No care in later life can restore the stamina of the body ill nourished, or unwisely nourished during those first seven years of life. With the joint family system there were children enough in the household, including those of the dependents, to make a society for the children, in which they learned unconsciously lessons of kindness, of courtesy, of gentle manners and refined speech, of little sacrifices born of love, of mutual helpfulness and mutual service.

With the narrowing of the home circle, the playing school is in many ways better and the children are happier in the merry games and the gay company of their little comrades. But the school must be well chosen, the teachers tender and helpful, songs, stories and play that exercises and trains the senses, the hand and the eye, and teaches graceful harmonious movements, are enough. From seven to fourteen are the years for training the memory and the emotions, for the stories of heroism and of virtue that inspire, drawn from the history of the Motherland, and great men and women; stories too of other countries; of all that can arouse enthusiasm and inspire to service.

Thus will the children have their minds and emotions so trained as to fit them to cross in safety the perilous bridge between childhood and youth. From fourteen to twenty-one is the time for hard mental study. By sixteen, the special capacities will have shown themselves, and will mark out the best avocation for the future life, and specialised Education may safely begin.

This is but the barest indication of the broad stages in the preparation for manhood and womanhood, the Ideal of the Student Order of the well regulated life. But the knell of popular education was struck in 1854, when Sir Charles Wood tried, and the Government supported, the singular experiment of teaching the people in a foreign tongue, with the result that after seventy years, 3.4 per cent of the people receive primary education. So we have three stages in Education in India in relation to the State: I. Lavish help from Rulers and complete liberty of Education, paid for by the wealthy and free to the poor, who, in exchange, served their teachers and performed household duties; II. Entire neglect for 97 years, with an interval of a lakh a year spent on it; III.

The Government English-speaking Schools, and Colleges, and later Universities with, of recent years, partial and grudging introduction of the vernaculars. How shall we apply the Indian Ideals to the salvation of Modern Education and Culture in India? That is the question which Indian Universities alone can solve, and before they can answer it, may, before they can even begin the task, the old relationship must be recreated between the State and the Universities.

Learning must again be inspired with the Ancient Ideals, and these will be embodied in new forms. And in order that these new forms shall be expressions of India's life, and not strait jackets to confine her, the old freedom must be restored to Education and Culture. Government should assign to educational and cultural institutions the material means for their support, gifts of land, grants of money for buildings, and for the necessary equipment, so that they may be able to give to the Nation the priceless assets of learned and skilled men and women of high character, to carry on the work in every department of national life.

Money given to Education by the Nation is not a gift, but an investment. It returns high interests to the Nation as well as power and happiness to the individual. Learned men produce literature which raises the Nation in the eyes of the world and, far more important, spreads knowledge over the earth, literature which ennobles and inspires not only contemporaries, but generations yet unborn.

Science makes discoveries which add to human knowledge, increase man's power over the forces on Nature, and—if it tread only righteous paths—will preserve, uplift any strengthen human life and human happiness. By Education and Culture of man's spiritual, intellectual, emotional and physical nature can he be lifted from the savage to the Sage and the Saint, can poverty be abolished, can society be made fraternal instead of barbarous, can crime, the fruit of ignorance, be gotten rid of, and international and social peace replace war and the strife of classes. Avidya is the mother of poverty, of sorrow, of misery.

It is the darkness which the Sun of Vidya must chase away. A generation of really educated people, with a proportion of the cultured, will change the face of India. Japan educated her people in forty years. As rapid as was the destruction may be the recovery, and each successive generation will show an improved result. Already Indian Ministers have made Primary Education free in seven Provinces and compulsory in three, compulsion to be introduced as rapidly as possible in the other four.

When India gains her own political freedom, may she be wise enough to restore freedom to Education and Culture, and, once more, the highest honour to Learning. After Freedom in the Educational and Cultural field is won, for it is not possible until this freedom is possessed, the very first thing must be the restoration of the Mother-tongues of India to their proper place in that field.

Nothing so denationalises a people as the imposition upon them of a foreign tongue, dominating their life and thought. When Germany, Russia and Austria rent Poland into three fragments, each banned the Polish tongue in the schools and imposed its own. Macaulay, with the most generous feeling and the most utter ignorance, urged the substitution of the English language, literature and civilization for those which he regarded as heathen and superstitious.

The Mother-tongues were despised, and a gulf was dug between the English-educated minority and the learned in the ancient Mother-language and the middle classes education in tongues derived from it. The free Universities will use the languages of the century throughout all schools and colleges, with English as a second language, and probably other tongues as well. So far, the Universities have given little culture; that has been gained by individuals for themselves. But free Universities will have curricula which shall give both Education and Culture. Students will, as of old, be surrounded with Beauty in the Schools, the Colleges, the Universities. The second basic difference between the Ancient System and the Modern English one, as imposed on India, is the absence of religious and moral education.

In Britain itself, the religion of the country and the morality based on it are taught in the schools as an integral part of education; lately, as Nonconformity and Free Thought spread, a conscience clause has been introduced exempting children, whose parents objected to the Anglican form of Christianity or to Christianity itself, from compulsory attendance at the religious services and lessons. But when the rule of the East India Company spread, and English Education was introduced into India, the Government schools dropped religious and moral teaching, since, on the one hand, a Christian Government could not teach heathen religions, and, on the other, as there were several religions in India, the Government must treat them all equally, and therefore remain neutral in regard to them.

Thus Indians must pay the taxes which keep up Government and other schools, and must further send their children to these, or to Missionary schools where an alien religion is taught, or open their own schools and teach any religion they belong to, Government giving them grants-in-aid. Modern Education in India had practically confined itself to the training of the mental and intellectual nature, and has ignored the unfolding of the spiritual nature, the evoking and training of the emotional nature, and, until lately, the development and training of the physical body to a high state of efficiency.

The result has been, in the older generations, the overstrain of the nervous system, the enfeebling of the physical health, the shortening of the period of vigorous maturity, often a sudden breakdown, or, at best, the premature appearance of debility and old age.

Further, the exclusive development of the intelligence and the neglect of the emotions has overstimulated the self-regarding instincts, and has largely destroyed the feeling of Social and National Dharma, of duty to Society and to the Nation; hence the decay of public spirit, of social service, of responsibility and of sacrifice for the common weal, which characterise the good *citizen* as distinguished from the good *man.*

WEB BASED OR VIRTUAL EDUCATION

Virtual education refers to instruction in a learning environment where teacher and student are separated by time or space, or both, and the teacher provides course content through course management applications, multimedia resources, the Internet, videoconferencing, etc. Students receive the content and communicate with the teacher via the same technologies. Virtual education is a term describing online education using the Internet. This term is primarily used in higher education where so-called Virtual Universities have been established.

Virtual courses – a synonym is online courses – are courses delivered on the Internet. "Virtual" is used here to characterize the fact that the course is not taught in a classroom face-to-face but through some substitute mode that can be associated with classroom teaching.

A virtual program (or a virtual course of studies) is a study program in which all courses or at least a significant portion of the courses are virtual courses.

Many virtual study programs are mainly text based, using HTML, PowerPoint, or PDF documents. Multimedia technologies have been investigated for many years and eventually found their way into practice. Today a wide spectrum of instruction modes is available, including the following:

- *Virtual Classroom:* A virtual classroom is a learning environment created in the virtual space. The objectives of a virtual classroom are to improve access to advanced educational experiences by allowing students and instructors to participate in remote learning communities using personal computers; and to improve the quality and effectiveness of education by using the computer to support a collaborative learning process. The explosion of the knowledge age has changed the context of what is learnt and how it is learnt – the concept of virtual classrooms is a manifestation of this knowledge revolution.
- *Hypertext courses:* Structured course material is used as in a conventional distance education program. However, all material is provided electronically and can be viewed with a browser. Hyperlinks connect text, multimedia parts and exercises in a meaningful way.
- *Video-based courses* are like face-to-face classroom courses, with a lecturer speaking and Powerpoint slides or online examples used for illustration. Video-streaming technologies is used. Students watch the video by means of freeware or plug-ins (e.g. Windows Media Player, RealPlayer).
- *Audio-based courses* are similar but instead of moving pictures only the sound track of the lecturer is provided. Often the course pages are enhanced with a text transcription of the lecture.
- *Animated courses:* Enriching text-oriented or audio-based course material by animations is generally a good way of making the content and its appearance more interesting. Animations are created using Macromedia Flash or similar technologies. These animations help understand key concepts and also allow for better retention of learning.

- *Web-supported textbook courses* are based on specific textbooks. Students read and reflect the chapters by themselves. Review questions, topics for discussion, exercises, case studies, etc. are given chapterwise on a website and discussed with the lecturer. Class meetings may be held to discuss matters in a chatroom, for example.
- *Peer-to-peer courses* are courses taught "on-demand" and without a prepared curriculum. A new field of online education has emerged in 2007 through new online education platforms.

Students in virtual education typically acquire knowledge in a uni-directional manner first (e.g. by studying a video, reading a textbook chapter). Subsequent discussions of problems, solving exercises, case studies, review questions, etc. help the students to understand better what they learned before. Electronic media like a discussion forum, chat room, voice mail, e-mail, etc. are often employed for communication. Homework assignments are normally submitted electronically, e.g. as an attachment to an e-mail. When help is needed, lecturers, tutors, or fellow students, or a help desk are available, just like in a real university. The difference is that all communication goes via electronic media. Most virtual study programs use an e-learning platform (learning management system – LMS) to administer students and courses and to provide learning content. Among them are Blackboard, WebCT, and Moodle.

- Second Life has recently become a virtual classroom for major colleges and universities, including Princeton, Rice University, University of Derby (UK), Vassar, the Open University (UK),. In 2007 Second Life started to be used for foreign language tuition. Both Second Life and real life language educators have begun to use the virtual world for language tuition. English (as a foreign language) has gained a presence through several schools, including the British Council], which has focused on the Teen Grid. Spain's language and cultural institute, the Instituto Cervantes has an island on Second Life. A list of educational projects (including some language schools) in Second Life can be found on the SimTeach site.
- WebEx is also increasingly used as an online learning platform and classroom for a diverse set of education providers such as Fox School of Business for Templer University, Grades Grow, Minnesota State Colleges and Universities, and Sachem. Webex is a Cisco Web Meetings and Collaboration Solution. The platform has worked for educational institutions because of real time collaboration using an interactive whiteboard, chat, and VOIP technology that allows audio and video sharing. In distance learning situations, while replacing the classroom with features, institutions have also looked for security features which

are inherently strong in a Cisco powered collaboration environment. The downside is that Webex is not a free platform.

GENERAL TRENDS IN ICT CAPACITY

The capacity for information and communications technology (ICT) has been growing exponentially over the last 10 to 15 years. Computers have become more powerful; satellite, fibre optic and wireless technology has increased transmission capacity; and software developments such as multimedia authoring systems have made it easier to create digital materials such as electronic games, computer simulations and educational materials. The increased capacity and supply has resulted in plunging prices for computer functionality, telephone services and software. Nowhere has the impact been greater than on the telephone industry. Long distance charges in North America and between major developed countries have dropped by 90% in the last 15 years.

Some analysts believe that the Internet will transform the pricing structure for telecommunications from charges by distance and time to a flat charge by capacity.

Thus a client will pay according to the size of the "pipe," not by the distance or length of time for which it is used. It should be noted that while the growth of the Internet is one of the major drivers for change, it is only one of several different but converging technologies that are resulting in widespread capacity growth.

The rapid growth in satellite and wireless technology, and in particular the growth in mobile phone technology, is resulting in some countries attempting to leap beyond previous or intermediate technologies. For instance, telephone capacity in China had been limited by lack of hard-wired networks. Mobile phone technology is allowing for a rapid expansion of telephone access in China. It is the last link to the end-user that has previously made hard-wired networks expensive to install on a widespread basis. Wireless technology avoids the need for hard-wired circuits into homes and offices.

However, there are several conditions surrounding this rapid growth in ICT, which can be summarised as follows:

- Despite the recent adjustments to world stock markets, capacity in ICT is forecast to continue to grow rapidly. For instance, while both Cisco and Nortel, the main providers of network hardware for the Internet, have revised downwards their growth forecasts this year, both anticipate continued growth of 15% or more annually for the next five years. This suggests that access, bandwidth and applications will all continue to increase into the foreseeable future.

- The technical capacity has far exceeded the capacity of governments, commercial organisations and educational communities to respond fully to the opportunities and challenges this rapid change has brought.
- ICT growth is not equal between different countries, between different socio-economic groups within countries or within different economic and social sectors of society. This inequality is often referred to as the digital divide: the gap between those with access and with the skills to use ICT appropriately, and those who do not have this capacity.
- Free trade and lower labour costs in developing countries in manufacturing and in resource- based industries such as agriculture and forestry have forced many former industrial nations, as well as some newly emerging "economic tigers" such as Malaysia and India, to move more aggressively into knowledgebased businesses heavily dependent on ICT technologies. Thus there has been a rapid growth in e-commerce, software development, hi-tech design (such as micro-processors, digital routers, computers and digital telephone switches), entertainment (video and computer games, film and television) and financial industries in developed countries. Such knowledge-based businesses both exploit the advantage of a highly educated workforce and, at the same time, drive the demand for an even *more* highly educated workforce. Much of this growth has occurred in smaller, more dynamic companies that avoid hierarchical manage-ment structures. The development of alternative organisational and management structures for the new knowledge-based industries is also relevant to virtual education, which is not only dependent on an extensive and reliable ICT infrastructure, but also requires a post-industrial approach to organisation and management.
- Technologies such as wireless and the Internet are disruptive in the sense that they bring about radical change to previously stable sectors. Companies such as Apple, Netscape, and amazon.com have helped revolutionise certain industry sectors, forcing previously dominant corporations either to radically change their operations or to close down. These same forces have the potential to bring about similar changes in the education sector.

Given the impact of ICT on many businesses and industries, it is not surprising that many commentators have seen education as being potentially revolutionised by it. Hence, there is great interest in virtual education, which is based on the idea of a widespread and significant application of ICT to the core activities of education. However, despite a great deal of hyperbole about the potential of ICT for education, real growth and change has been slow and marginal. This is due not only to lack of vision or commitment by educators and policy-makers; there are significant structural and cultural barriers or restrictions that have slowed the potential for change in education compared with other sectors. These issues are discussed below, first looking at some case studies, then examining what kinds of technology are being used and why. Following that is a brief examination of some key emerging technologies and how they may have an impact on the education sector.

Brief Case Studies of Major ICT Applications in Education and Training

As there is a very wide range of different applications of ICT in education, any choice of case studies is somewhat arbitrary. For example, the Open University in the U.K. (UKOU) has been using communications technologies such as broadcast television and radio for educational delivery since the early 1970s, and in recent years it has also moved into online programming. However, print still remains the core delivery medium for the UKOU, even though it has expanded its programming into Europe and other areas of the world.

Similar "dedicated" national open universities, such as Indira Ghandi National Open University in India, the Chinese Central and State Radio and Television Universities, and the Korean National Open University, have operated for many years, primarily on a mix of broadcasting and printed materials. Other dedicated open universities, such as Athabasca University in Canada, have moved more aggressively into Web-based delivery.

At the same time, many institutions that offer both campus-based and distance programmes, such as the University of British Columbia (UBC) in Canada, Charles Sturt University in Australia, and Penn State University in the U.S.A., have begun to move their former print-based distance education courses online. For instance, UBC has over half of its 110 distance courses now online. In Korea and several other countries, campusbased universities are forming consortia for the delivery of online distance learning programmes. However, the most dramatic change in the use of ICT has not been in the traditional distance education market.

The biggest impact has been on campus-based teaching and in the private sector training market. John Chambers, the CEO of Cisco, one of the world's largest ICT companies, has stated that "the next big killer application for the Internet is going to be education". Merrill Lynch has described e-learning (the combination of the corporate learning and higher education markets) as a U.S.$18 billion market by 2003, compared with $2.3 billion market in 2000.

Here are just a few examples:

- WebCT, a course authoring software platform, has over three million student licences in over 1500 institutions, and it is growing at a rate faster than 500,000 new student licences a year. However, WebCT estimates that over 80% of the use of its software is to support classroom teaching rather than "pure" distance education. Thus we are seeing the emergence of "mixed mode" teaching, combining face-to-face instruction with online learning. One of the best examples of this is the University of Central Florida, which has an extensive programme of Web-based courses available both to on-campus and off-campus students.

- A number of universities have formed consortia and partnerships with the private sector to commercially exploit their e-learning initiatives. Universitas 21 is a network of 17 universities in mainly Commonwealth countries, including Nottingham University, University of Edinburgh, University of Melbourne, Hong Kong University, National University of Singapore and the University of British Columbia. Thomson Publishing and Universitas 21 have announced a partnership to found an "e-university." In 1990 there were 48 million people in higher education by 2025, Thomson Publishing estimates that there will be 160 million.
- UNext, a U.S. company, has established an e-university called Cardean that adapts its teaching material from that of the universities of Columbia, Stanford, Chicago, Carnegie Mellon and the London School of Economics. Degrees are awarded under the Cardean name, endorsed by the state of Illinois.
- The London School of Economics is also more deeply involved in an e-learning venture called Fathom launched last year as a global online library linking institutions such as the New York Public Library, the British Library, the Smithsonian, the Cambridge University Press and the LSE. Fathom's partners participate in the running of the service, and all must approve materials before they are published on the site.
- The U.K. government has established one of the most ambitious joint e-university projects. It has put the Higher Education Funding Council for England (HEFCE) in charge of attempts to create an e-university with a budget of £400 million, half of which will be public sector money. At first this was intended to be an exclusive arrangement, with universities bidding to become members. But now all U.K. universities are allowed to hold shares. The only human contact in the core programme will be with "navigators" — advisers who will help new students to select courses. Only those students who attend summer schools or pay for additional tutorial support will receive face-to-face tuition.

- Oxford University is linking with Stanford, Yale and Princeton to create an online college for alumni. The idea is to provide lifelong learning and support to some of the world's policy-makers and business leaders. Cambridge is exploring virtual learning in its £83 million government-backed link-up with the Massachusetts Institute of Technology (MIT). The project is known as the "bridge of minds" and is principally intended to foster commercial spin-offs from university research. It also has potential to become an euniversity.
- A U.S. company, Hungry Minds has contracted with Michigan State University, New York University, Penn State University, Rochester Institute of Technology, University of California Berkeley Extension and University of California at Los Angeles Extension to provide a range of e-learning services. These vary from providing a common Web portal for their courses, to supporting the development of online certificates by these institutions.
- Since 1996, the University of British Columbia (UBC) has been offering a fully online graduate level programme on technologybased distributed learning in partnership with the Monterrey Institute of Technology in Mexico. The core of the programme is on the Web, but UBC also participates by video-conference link from Vancouver to Monterrey in satellite television broadcasts to Monterrey Tech's 29 campuses across Latin America. Monterrey incorporates the five courses in its own master's degree in Educational Technology and has the rights to offer the five courses throughout Latin America. UBC has made the five courses available to both its own mainly campus- based master's students as well as to the rest of the world as a post-graduate certificate programme. To date, it has recruited students from more than 30 different countries. This programme runs on a self-financing basis entirely from student fees. As a result of this experiment, UBC and Monterrey plan to offer a joint master's degree in educational technology in both Spanish and English on a global basis from January 2002.

In the private sector training area, there has been a rapid expansion of "corporate universities," primarily in-house training organisations in the larger multinational companies making use of videoconferencing and the Internet. In parallel, there has been the development of a whole new industry of online training contractors who provide online training services to medium- and smaller-sized companies. Almost two-thirds of Merrill Lynch's estimated $18 billion market is in this private sector training area.

A number of private, for-profit organisations are trying to bridge the gap between centralised distribution (such as foreign Web courses or satellite TV) and local support. For instance, TeltecGlobal is a "business services aggregator" offering corporations and governments "a one-stop, turn-key solution for access to 21st century technology, services and education." The TeltecGlobal Center of Influence is a "last mile" strategy that provides customers with products and Web-enabled services available through membership in their Community and Business Centers located in developing countries.

TeltecGlobal Community Centers work in conjunction with multinational corporate sponsors, designing an offering of products and services "to meet both community needs and corporate objectives." Business Centers license operations to local entrepreneurs and "global governmentbacked entities" in emerging markets. As with the Community Centers, TeltecGlobal works with the licensees to tailor the products and services to market needs and licensees' goals. TeletecGlobal is a good example of the increasing synergy between technology, education and business. These developments are not so different from the chaotic, diverse and excessively optimistic developments in the dot.com business sector. Indeed, especially in the U.S.A., e-learning is seen as just another branch of e-business. However, a more careful analysis of these developments will indicate that there are many barriers and challenges that need to be overcome before such initiatives can operate on a sustainable basis.

NEED OF VALUE EDUCATION IN 21 CENTURY

Education is a methodical effort towards learning basic facts about humanity. And the core idea behind **value education** is to cultivate essential values in the students so that the civilization that teaches us to manage complexities can be sustained and further developed. It begins at home and it is continued in schools. Everyone accepts certain things in his/her life through various mediums like society or government.

Value education is important to help everyone in improving the value system that he/she holds and put them to use. Once, everyone has understood their values in life they can examine and control the various choices they make in their life. One has to frequently uphold the various types of values in his life such as cultural values, universal values, personal values and social values.

Thus, **value education** is always essential to shape one's life and to give him an opportunity of performing himself on the global stage. The need for **value education** among the parents, children, teachers etc., is constantly increasing as we continue to witness increasing violent activities, behavioural disorder, lack of unity in the society etc.

The family system in India has a long tradition of imparting value education. But with the progress of modernity and fast changing role of the parents it has not been very easy for the parents to impart relevant values in their wards. Therefore many institutes today conduct various value education programs that are addressed to rising problems of the modern society. These programs concentrate on the development of the children, young adults etc. focusing on areas like happiness, humility, cooperation, honesty, simplicity, love, unity, peace etc.

Values

Values are those things that really matter to each of us... the ideas and beliefs we hold as special. Caring for others, for example, is a value; so is the freedom to express our opinions.

Most of us learned our values-or morals, if you prefer-at home, at church or synagogue, at school. But, where are our children learning their values? Maybe from parents, teachers and religious leaders, but society has changed. Too often young people today are most influenced by what they see and hear on television or on the street.

For this reason, the Boy Scouts of America-the nation's largest youth development organization-introduced new tools to help young people-from Cub Scouts through Exploring-develop positive values while learning to make ethical decisions.

The Scout Oath and Law express a well-defined code of ethical and moral conduct. If you think about it, you'll see that these abstract ideas-trustworthy, loyal, helpful, friendly, courteous, kind, obedient, cheerful, thrifty, brave, clean, reverent-can become very concrete goals for young people.

No, these ideas aren't "new and improved." They've been around since the very beginning of Scouting. These new tools provide a means for teaching today's young people how to apply these abstract ideas in everyday situations. Scouting is designed to keep young people busy and involved in all sorts of projects. Scouts learn by doing, and learning values is as action-oriented as other Scouting programs. And all are designed to help young people acquire the values that they will strive toward well beyond their Scouting years.

After years of talk about the "moral decay" in just about every area of American life, our society seems to be turning back to the traditional values that guided this nation to greatness. To pass these values on to our children through Scouting relies upon three components: caring adults, age-appropriate and purposeful activities, and meaningful roles in the community

Leadership training and the literature of Scouting have been revised to help caring adults become better, more effective Scout leaders... to recognize that young people develop physically, mentally, socially-and, yes, ethically-at different rates... to identify community service projects that drive home the message that young people, by interacting with people in their community, can have a positive impact on the world around them. Building upon Scouting's long history of youth development, makes this heritage even more relevant in meeting the needs of contemporary youth.

In ethics, **value** is a property of objects, including physical objects as well as abstract objects (e.g. actions), representing their degree of importance.

Ethic value denotes something's degree of importance, with the aim of determining what action or life is best to do or live, or at least attempt to describe the value of different actions. It may be described as treating actions themselves as abstract objects, putting value to them. It deals with right conduct and good life, in the sense that a highly, or at least relatively highly, valuable action or may be regarded as ethic "good" (adjective sense), and an action of low, or at least relatively low, value may be regarded as "bad". What makes an action valuable may in turn depend on the ethic values of the objects it increases, decreases or alters.

An object with "ethic value" may be termed an "ethic or philosophic good" **personal and cultural value** is a relative ethic value, an assumption upon which implementation can be extrapolated. A *value system* is a set of consistent values and measures that are not true. A*principle value* is a foundation upon which other values and measures of integrity are based. Values are considered subjective, vary across people and cultures and are in many ways aligned with belief and belief systems. Types of values include ethical/moral values, doctrinal/ideological (religious, political) values, social values, and aesthetic values. It is debated whether some values are intrinsic.

DIFFERING GLOBAL PERSPECTIVES

Perhaps the most important step in understanding and incorporating global education in classrooms and communities is to understand and relate to the themes of global awareness as presented by experts in the field. Hanvey (1976), one of the first scholarly experts to give a comprehensive definition of the concept "global awareness", proposes five dimensions that prepare students to achieve global awareness. These include perspective consciousness, state-of-the-planet awareness, cross-cultural awareness, knowledge of global dynamics, and awareness of human choices. Haavenson, Savukova, and Mason (1998/99) conducted their research on United States and Russian perspectives on teacher education reform and global education and found that these dimensions form the first level known as attitude formation upon which global education can be implemented. The second level is the development of cognition skills and the third level is an integrated view of the world. An explanation of Hanvey's five dimensions, paraphrased by Kirkwood (2001) and Haavenson et al. (1998/99) is provided below and an explanation of the other two levels will be identified.

Perspective Consciousness

Perspective consciousness refers to an awareness of and appreciation for other images of the world and that a person's worldview is neither universally shared, nor necessarily right, yet may be profoundly different. It is the realization that an individual's worldview is both a matter of conscious opinions and ideas and more importantly to subconscious evaluations, conceptions and unexamined assumptions.

Perspectives are shaped by ethnic, religious, differences in age, sex, and social status, among many other factors. These differences, as stated by Haavenson et al. (1998/99), "have been one of the main causes of conflict and confrontation in the history of mankind" (p.38).

The authors go on to say that, "It is important to teach students to look upon a certain phenomenon or event from different perspectives so as to encourage respect and appreciation for beliefs, customs, and values different from their own" (p. 38). It is not only about racial and cultural differences, instead, a pluralistic view needs to be taken when looking at global perspectives.

State-of-the-planet Awareness

State-of-the-planet awareness requires comprehension of prevailing world conditions, developments, trends, and problems that are confronting the world community. It includes an in-depth understanding of global issues such as population growth, migrations, economic disparities, depletion of resources, and international conflicts, that require global learners to be aware of the world around them. Children need to be made aware that what affects the world affects them as well. In elementary school, students can be taught to make decisions about ways to prevent disaster by studying the consequences of environmental illiteracy.

Cross-cultural Awareness

This dimension includes the diversity of ideas and practices in human societies and how these ideas and practices are found in human societies around the world, including concepts of how others might view one's own society as perceived from other vantage points. According to Hanvey (1976), this dimension is the most difficult to attain most likely because it refers to the highest level of global cognition. The misconception about cross-cultural awareness is that people consider it no more than a set of stereotypes that do more harm than good as superficial knowledge engenders prejudice.

An effective way to promote cross-cultural awareness, as explained by Haavenson et al. (1998/99), is by showing videos and then having discussions with students about these films to help them in separating stereotypical views from those that are more authentic.

Knowledge of Global Dynamics

Knowledge of global dynamics refers to an understanding of the world as an interconnected system of complex traits and mechanisms and unanticipated consequences. A high level of sophistication on the part of the student is required because understanding these processes is difficult to achieve due to the unanticipated effects on the human condition. It includes a consciousness of global change and cannot be acquired through mass media. Haavenson et al. (1998/99) explain that "[s]tudents learn to identify subtle cause-effect relationships, anticipate side effects, model processes and make decisions about eliminating or altering undesirable consequences". Students may be asked to create webs of the factors influencing the issue, to suggest feasible solutions, and to foresee possible side effects of such actions.

Awareness of Human Choices

Hanvey (1976) challenges global thinkers to realize the problems of choice confronting individuals and nations as consciousness and knowledge of global systems expand. It is related to global dynamics in such a way that it focuses on making choices and develops a sense of responsibility for making decisions made which affect future generations. It also includes an awareness of the interconnectedness of individual, national, and international settings. It fosters a sense of responsible citizenship on the local and global levels. Students may be introduced to alternatives on thought and behaviour by looking at relationships and interactions between man and the world. Students are asked to account for their choices and are taught to be tolerant toward the view of others.

In their study with teachers, Haavenson et al. (1998/99) posits that the second level of global education implementation is cognition focused. This means that life demands both a thorough knowledge of a domain combined with a broad perspective of the world. This is similar to the 'interconnections' theme that Werner & Case (1997) identify and develop which explores both the international and inter-system linkages and conclude that we live in an interconnected world. Therefore students must be encouraged to see the different ways in which one situation is influenced by and influences others.

Further exploration in the topic is explained by Haavenson et al (1998/99) that the brain often searches for common patterns and relationships and seeks to connect new knowledge with prior experiences that result in the fact that cognition operates in all concepts. The traditional approach of filling the minds with facts and information that students are simply asked to memorize and reproduce does nothing to promote global awareness and teachers must keep this in mind when working to plan curriculum. Instead, students need experience in critical thinking, in taking part in cross-cultural experiences, and to make decisions and substantiate them. In the study by Haavenson et al. (1998/99), students are taught to think for themselves and to be able to stand their ground. The authors advise that the atmosphere created by the teacher is very important.

The third level of global education implementation is an integrated view of the world as explained by Haavenson et. al. (1998/99). They state that "the third level aims to create a specific picture of the world where geographical, physical and linguistic features all fit into a complex pattern". This means that all discipline-focused world perspectives need to overlap due to the interdependence of facts, events, and phenomena. For instance, university interdisciplinary courses may be the most effective way to create a cross-disciplinary perspective.

Global Awareness Elements

Case (1993) identifies five key substantive elements that keep people informed of a range of global topics. The first element describes the universal values and cultural practices, and the second includes global interconnections, which refers to the study of the workings of the four major interactive global systems: economic, political, ecological, and technological.

The third presents worldwide concerns and conditions such as development and peace issues while the fourth forms the origins and past patterns of worldwide affairs such as global history and geography. The last presents alternative future directions in worldwide affairs. In addition to these substantive elements, he proposes perceptual elements that should be addressed which include open mindedness, resistance to stereotyping, anticipation of complexity, empathy, and nonchauvinism.

Kirkwood (2001) analyses Case's elements and explains that the substantive elements listed above "includes the objects of global education that incorporate the contemporary events, conditions and locations in the world that Hanvey (1976) addresses within the state of the planet awareness dimension".

The perceptual elements focus on the development of world mindedness and empathy and resistance to prejudicial thinking as well as stereotyping and cross-cultural knowledge. These elements are similar to the Hanvey dimensions of perspective consciousness and cross-cultural awareness.

Case (1993) and Hanvey (1976) provide similar definitions for global awareness even though the terminology they use is different. Merry M. Merryfield, one of the leading scholars in the field of global education, combines the definitions of other scholars and provides us with a current framework in this field today.

Kirkwood (2001) lists Merryfield's eight elements which include: human beliefs and values global systems, global issues and problems, cross-cultural understanding, awareness of human choices, global history, acquisition of indigenous knowledge, and development of analytical, evaluative, and participatory skills. Kirkwood (2001) concludes that "Merryfield's work contributes significantly in reducing, if not eliminating, the definitional ambiguities that still linger in the field".

New Understandings of Global Awareness

Kirkwood (2001) presents another dimension to the definition of global education that Lamy (1987) identifies as the acquisition of knowledge transmitted by indigenous people. He concludes that a global education must include knowledge about the contributions of native people who are representing the views of their world. In his words, "The teaching of historical and contemporary events must be balanced by listening to indigenous voices".

To provide further elaboration in regards to listening to Indigenous voices, Battiste and Henderson (2000) think of globalization as a new threatening transformation that is emerging. In the introduction of their book, they state that, "Globalization with its cognitive and linguistic imperialism is the modern force that is taking our heritages, knowledge, and creativity".

For Indigenous people it is not just physical survival that concerns them; it is an issue of "maintaining Indigenous worldviews, languages, and environments". It is ironic that the world looks to Indigenous people for help in order to solve the world's crisis that its worldview has created. Battiste and Henderson (2000) state that "in view of the history of relations between the colonizers and the colonized, this is an extraordinarily bold request."

The work of David Selby and Graham Pike has brought new understandings in regards to ecological awareness or 'state of the planet awareness' as outlined by Hanvey. They have been influential in the global education field in the 1980's and are known nation wide. Influenced mainly by Richardson and Hanvey in the 1970's they have worked mainly with secondary schools. Hicks (2003) in his review of global education discusses the work of Selby & Pike.

He explains that in 1988 they further developed the conceptual map of the field and highlighted what they called 'the four dimensions of globality'. These dimensions make up the core elements of global education. The first is 'issues dimension', which embraces five major problem areas and solutions to them: inequality/equality, injustice/justice, conflict/peace; environmental damage/care; alienation/participation.

The second is 'spatial dimension' which emphasizes exploration of the local-global connections that exist in relation to these issues, including the nature of both interdependency and dependency. The third is 'temporal dimension' that emphasizes exploration of the interconnections that exist between past, present, and future in relation to such issues and in particular scenarios of preferred futures. The fourth is the 'process dimension' that emphasizes a participatory and experiential pedagogy which explores differing value perspectives and leads to politically aware local-global citizenship. Selby and Pike then relate this to both individual subjects in the curriculum and whole-school case studies.

Hicks (2003) further explains that each of these four elements needs to be present before one can claim to be involved in global education. Both Selby & Pike have written extensively on the importance of ecological thinking in global education and this is evident within the four-dimensional model that they propose for global education. It needs to be stressed that the environmental health of the world is just as important as taking care of all humanity and that the two must work together simultaneously. For example when explaining the 'spatial dimension' Selby (1998) writes that "this dimension also concerns the cycles and systems of nature and the relationships between human society and the environment".

Hicks (2003) explains that the 'temporal dimension' is a futures perspective that "looks at how global issues affect and are affected by interrelationships between past, present and future". He goes on to say that "this works to help young people think more critically and creatively about the future, especially in relation to creating more just and sustainable futures".

EDUCATIONAL SCENARIO IN INDIA TODAY

More and more parents, therefore, were turning towards private schools even in villages and wayside slums. The PROBE team discovered that private schools, even in remote villages, were very active in teaching and getting good results for their pupils. In the private schools the teachers did not have job security and consequently, their continuation depended on their performance, watched both by the owners who wanted a fair name for their schools and the parents, who could shop around for a school with a good name anyway. That motivates the teachers to work hard and put their heart and soul into teaching. Those village schools charged very reasonable sum of 35-50 rupees per month; in the city slums the going rate was 60-100 rupees per month.

These private schools were more popular because they all taught English, in addition. In the slums around Charminar in Hyderabad alone there were as many as 500 such schools belonging to a single Federation, serving predominantly the poorer sections of society like the Rickshaw pullers, daily wage earners, vegetable sellers, fisher women, and the like. They charged very modest fee of around a thousand rupees an year. None of them depended on government subsidy! These schools also had an altruistic motive in reserving a quarter of their seats for the poorest of the poor in the locality giving them away free. Similar experiments were going on in many other developing countries like Thailand, Columbia, Tanzania, and Chile. This method could be easily replicated elsewhere. In fact, Prof. Tooley recommends many of these methods to the inner city area schools that starve for want of funds and good teachers.

This self-help idea was the one that prompted the first ever private medical college in India in Manipal by a thinking man, Late Dr. T. M. A. Pai, who wanted that the motivated students should be provided with an opportunity to pursue their interest, assisted by their parents chipping in their lot to sustain the institutions. The Manipal Academy of Higher Education, a Deemed to be University, is the result of his initial effort. Independent surveys have given us very high rating not only in India but abroad. We have students from thirty English speaking countries. We have been assessed as one of top four colleges in medical education.

Institutions do not depend for their growth on either the government or other owners. Educational institutions get their name and fame because of the men and women who struggle to keep up the highest academic standards as also the ethical values. Our founder's motto was to nurture the best of teachers. Having been in these institutions for forty years I could vouch for that truth. Universities should be proud of their faculty and not their brick and mortar or equipment as much. Our powers-that-be do not seem to realize this naked truth.

It is only in our country that private effort in education is being looked down upon and treated badly. People outside governmental control have founded some of the great institutions all over. A banker, like Dr. TMA Pai, founded the University of Edinburgh, way back in the eighteenth century. That was the time of the Scottish enlightenment when Edinburgh was considered to be the Athens of the north. Except the University of Georgia Medical School, most of the American medical institutions of repute are in the private sector. Harvard, Yale, Stanford, Johns Hopkins, Mayo Clinic, Cleveland Clinic and most of the others belong to that class. The Guy brothers, business tycoons of those days, founded the famous Guy's Hospital Medical School in London. They are not untouchables in their countries, but are venerated very much as centres of excellence and repute.

This pioneering Manipal experiment, the first of its kind in India, unfortunately has been badly replicated by many others to make it into a big "business" in higher education. These kind of unscrupulous methods are being abetted and nurtured by politicians and their goons for their own benefit. That is no reason why the original idea of Dr. T.M.A.Pai should be found fault with by our thinkers. Many of the later institutions do not have even the bare minimum necessities, not to speak of the all-important faculty. It would be shocking to know that a sizable percentage of them thrive only on visiting faculty, who rarely visit. People in authority would overlook all these so long as their machinery is well oiled and greased! Thanks to the munificence of our greedy powers-that-be, such institutions thrive better in India. It is only the honest and the meritorious that suffer in this environment.

What is the need of the hour? Many of our journalists with the holier-than-though attitude towards private efforts at higher education and many of our armchair intellectuals who live in ivory towers, having no touch with reality and without any personal experience in the field, think that education, outside the government setting, is getting commercialized in India. They feel that this is the greatest sin and should be curbed at any cost. We would be happy to host them here to have first hand knowledge of the trials and tribulations of running excellent educational institutions.

Similar arguments were put forward for "socialism" of the Russian variety in the '50s, despite the fact the Mahatma Gandhi had strongly advocated the cottage industry and village development as the need of the hour. Now even the champions of the mega things in development are convinced that "small is beautiful". The country is suffering from the fall out of that sin at an enormous cost to the people. Governments got involved in every aspect of the common man's life, starting from transport, electricity, water supply, food distribution, industry, health care, hotels, and what have you. None of them function properly even after half a century! It has now dawned on the politicians that they should get out of these as soon as possible. Education is another field where the government has miserably failed. Sooner they realize this the better for our future generations.

We must nurture and develop private schools and even professional colleges that keep up excellent standards of education. There should be no compromise on standards. Let us learn the lessons of the PROBE study. Let private institutions depend on their excellence for their very existence! Allow them a level playground without throttling them. Let there be an independent accreditation body of independent people of integrity that periodically publicizes the standards for students and parents to know.

Let there be survival of the fittest without any outside agency bringing in unreasonable restrictions. Let the buyer, the well-informed student, take the pick. The bad institutions would die a natural death in the process. That is the exact way how American medical education was cleaned of the unscrupulous medical schools of which there were more than two hundred in the fifties. Flexner Committee, an independent body, rated all of them. Their findings were made public by the government repeatedly in national dailies over a period of a month or so. Only seventy-two colleges survived, as the rest died a natural death for want of student aspirants to fill their seats. No amount of regulations and rules would work that effectively, since the crooked owners could twist many of the latter.

This reminds me of a lecture given by Mrs. Margaret Thatcher, the then Prime Minister of Britain, in one of the conferences on "quality control". She got up to speak and started by narrating her own experience, as the grocer's daughter, at her father's shop, when she was young. "Whenever the vegetables we sold were good the customers came back; when the vegetables were bad the vegetables came back and the customers went away to other shops! That is quality control." How true! The institutions should survive on their merit alone and not on rules and controls in the open market. That would be healthy for higher educational institutions as well. The really bad institutions will die a natural death under those conditions, as no student would want to go there.

The private institutions should be allowed to grow without the usual constraints of the "licence-raj" let loose on them both by the politicians in the government and their henchmen. The above categories do not want to loosen their grip on the private institutions. The corrupt people's bread and butter are these institutions. That is precisely why they want to have their complicated regulations and rules to have an absolute control over private effort at education. Our biggest curse is this "license-raj" system. The root of all corruption starts there. The more rules that the government puts forward, the better for the greedy in power. Every single rule is an opportunity for corruption.

This country, or for that matter, even the developed countries, would not be able to subsidize primary education, leave alone higher education, in the new millennium. The recipient should pay for higher education. If he is poor or hails from an economically backward community, the government could help him raise a loan from the Educational Development Bank to be repaid only after he gets a job and the interest thereon could be waived or reduced depending on his economic status.

Most governments would not be able to service the interest on their foreign borrowings in the next decade even with all their revenue collection. Many of them would not have the money to pay their staff! In that situation education could only be in the private sector. Our netas should not be under the delusion, that by selling the idea to the gullible public that commercialization in education is bad for the country, they feel that they could continue to use this "milch cow" to fill their coffers for elections and also to hoard money for their progeny. They seem to forget that they can not take the hoarded money with them when they go to meet their maker.

Even the famous "Unnikrishnan" judgement of the Supreme Court subscribes to this view that private education is bad in principle. I am sure with the present scenario one could go back to the Supreme Court for a revision of that judgement in the background of the wisdom from the PROBE investigations. With more evidence accumulated against the view, our adversarial system of justice might agree with the new wisdom.

As of now the private institutions face great hostility from the government. They are viewed with suspicion as thieves. While there are thieves in every field of human endeavour, there are also excellent institutions run on ethical principles. Of course, there is no free lunch anywhere. The private institutions should be allowed to get back a reasonable return on their investments and also be able to break even to sustain themselves. A logical financial assessment of the needs vis-à-vis their expenditure could be made by an independent body whose members' credentials could be bared for public scrutiny before being appointed, but the committee should not have any kind of restrictions, either from the governments or the watchdog bodies. I am sure there are many in this country that are honest and ethical. In fact, they are in the majority, but they are a silent majority not noticed by our media and also the merit award giving bodies of the country! Let them be given at least this thankless job of overseeing private initiative in educating our masses in higher studies.

Many of our leaders in the government behave as if they belong to a higher race. I am reminded of what French President Giscard d'Estaing wrote in his memoirs in 1981. "In my country" he said, "there is the idea that those who govern belong to another race."

Earlier we change the set up and clean up the stables the better for the future generation. With more than five hundred million young men and women looking for higher education in the country in the next fifty years, we would create chaos in the near future unless we set our house in order. There is no desirable pattern of non-governmental educational efforts in higher education prescribed so far. Consequently, the unscrupulous would want to commercialize education. Unfortunately, they are the ones that get all the perks from the powers-that-be as they supply the needs of the latter. Those who want to be honest and authentic get all sorts of hurdles put on their way.

Every rule is made to be used to either make money for the rulers and their ilk or for having some control over the hapless organizations in the field of higher education. A survey of the present facilities for higher education in the existing set up would throw up worse things than what was brought out by PROBE team in primary education.

While the government set-ups have very little infrastructure, the private ones are made to achieve the impossible. I get a feeling that there must be a deep-rooted conspiracy to keep up the governmental hegemony in education for the politician and the bureaucrats to make money perpetually. This reminds me of the nice study published from the Cato Institue in Washington edited by Doug Bandow and Ian Vasquez entitled Perpetuating Poverty, wherein the authors have systematically shown how the World Bank and the International Monetary Fund have promoted poverty in the developing world for the ultimate good of the big powers, so that they could keep the former under their thumb perpetually. Writing a commentary on the book, Melvyn B. Krauss, professor of economics at the NewYork University, has this to say: "This book destroys the myth that the multilateral aid agencies are forces for good and shows they are instead a major fraud perpetuating poverty in the developing countries.............The editors are to be congratulated for editing this precise, cold-eyed, and extremely important collection of essays that merits a wide audience."

Bibliography

Adelman, C.: *The Politics and Ethics of Evaluation*, London, Croom Helm, 1984.

Aggarwal, Santosh : *Three Language Formula: An Educational Problem*, New Delhi, Sian, 1991.

Allen, G. : *New Education: American Opinion*, May issue, 1971.

Altbach, Philip G. and Gail Kelly : *New Approaches to Distance Education*, Chicago, The University of Chicago Press, 1986.

Beals, R. L.: *Politics of Social Research*, Chicago, Transaction Books/Aldine, 2005.

Chaklader, Snehamoy : *Sociolinguistics: A Guide to Language Problems in India*, New Delhi, Mittal, 1990.

Cole, Robert A.,: *Issues in Web-based pedagogy: a critical primer*, Westport, Conn.: Greenwood Press, 2000.

Connie L., Rosa Cintrón: *Building a Working Policy for Distance Education*, San Francisco, Jossey-Bass, 1997

Donaldson, Gordon A. : *Cultivating Leadership in Schools*, New York, College Press, 2001.

Eisenstadt, Marc: *The Knowledge Web: Learning and Collaborating on the Net*, London, K. Page; Sterling, VA: Stylus Pub., 2000

Elliott, J. : *Action Research for Educational Change*, Milton Keynes, Open University, 1991.

Evers, C. & Lakomski, G. : *Knowing Educational Administration*, Oxford, Pergamon, 1991.

Feenberg, A.: *Mindweave: Communication, Computers and Distance Education*, Elmsford, New York: Pergamon Press, 1989.

Fisher, William C. : *Working Toward Sustainable Development: The Narmada Dam Project*, Armonk, New York, 1994.

Goodlad, John I. : *Educational Renewal: Better Teachers, Better Schools*, San Francisco, Jossey-Bass, 1994.

Gordon A. : *Cultivating Leadership in Schools*, New York, College Press, 2001.

Grace, G. : *School Leadership: Beyond Educational Management*, London, Falmer, 1995.

Greg Kearsley: *Distance Education : A Systems View*, Wadsworth Pub Co., 1996.

Gulley, H. E.: *Discussion, Conference, and Group Process*, New York: Holt, Rinehart and Winston, 1968.

Hallenger, P.: *Problem-based Learning for Educational Administrators*, Oregon, ERIC Clearinghouse on Educational Management, 1992.

Harasim, L.: *Online Education — Perspectives on a New Environment*, New York, NY: Praeger Publishers, 1990.

Harry, Keith: *Higher Education through Open and Distance Learning*, New York: Routledge, 1999.

Jeffs, T. and Smith, M. : *Using Informal Education*, Milton Keynes, Open University Press, 1990.

Keltner, J.: *Group Discussion Processes*, Westport, Greenwood Press, Publishers, 1974.

Lovett, T. : *Distance Education, Community Development and the Working Class*, London, Ward Lock, 1975.

Lovett, T. : *Distance Education, Community Development and the Working Class*, London, Ward Lock, 1975.

McCarty, Willard: *Humanities Computing Yearbook*, Oxford: Clarendon Press, 1988.

McGivney, V. : *Informal Learning in the Community, A Trigger for Change and Development*, Leicester, NIACE, 1999.

Miller, D. C. : *Handbook of research design and social measurement*. New York: David Mc Kay Company, 1977.

Miller, N. : *Working with Experience: Animating Learning*, London, Routledge, 1997.

Moore, Michael G. and Greg Kearsley: *Distance Education : A Systems View*, Wadsworth Pub Co., 1996.

Naik, J.P.: *The Education Commission and After*, Allied Publishers (Pvt.) Ltd., New Delhi, 1982.

Novotny, Jeanne: *Distance Education in Nursing*, New York, Springer, 2000.

Nuna, S.C. : *Distance Education and Development*, NIEPA, New Delhi, 1987.

Passow, A.H. : *Education of the Disadvantaged*, New York, Holt, 1967.

Rogers, A. : *Adults Learning for Development*, London, Cassell, 1992.

Simkins, T. : *Non-formal Education and Development*, Manchester, Manchester University, 1977.

Simkins, T. : *Non-formal Education and Development*, Manchester, Manchester University, 1977.

Smith, M. : *Using Informal Education*, Milton Keynes, Open University Press, 1990.

Stenhouse, L. : *Authority, Education and Emancipation*, London, Heinemann, 1983.

Stewart, J. : *Management for the Public Domain*, New York, St, Martin's Press, 1994.

Stoner, W.S. : *Natural Education*, Indianapolis, Bobbs Merrill, 1914.

Thomas, Edward J. : *The Life of Buddha as Legend and History*, London, Routledge and Kegan Paul, 1927.

White, P. : *Civic Virtues and Public Schooling: Educating Citizens for a Democratic Society*, New York, Teachers College, 1996.

Index

❑❑❑